MATHEMATICS OF SHAPES
AND APPLICATIONS

LECTURE NOTES SERIES
Institute for Mathematical Sciences, National University of Singapore

Series Editors: Chitat Chong and Adrian Röllin
Institute for Mathematical Sciences
National University of Singapore

ISSN: 1793-0758

*For the complete list of titles in this series, please go to
http://www.worldscientific.com/series/LNIMSNUS

Lecture Notes Series, Institute for Mathematical Sciences, National University of Singapore   Vol. 37

MATHEMATICS OF SHAPES AND APPLICATIONS

Editors

Sergey Kushnarev
Singapore University of Technology and Design, Singapore

Anqi Qiu
National University of Singapore, Singapore

Laurent Younes
Johns Hopkins University, USA

World Scientific

NEW JERSEY · LONDON · SINGAPORE · BEIJING · SHANGHAI · HONG KONG · TAIPEI · CHENNAI · TOKYO

Published by

World Scientific Publishing Co. Pte. Ltd.

5 Toh Tuck Link, Singapore 596224

USA office: 27 Warren Street, Suite 401-402, Hackensack, NJ 07601

UK office: 57 Shelton Street, Covent Garden, London WC2H 9HE

Library of Congress Cataloging-in-Publication Data

Names: Kushnarev, Sergey, editor. | Qiu, Anqi, editor. | Younes, Laurent, 1963–   editor.
Title: Mathematics of shapes and applications / editors, Sergey Kushnarev,
    Anqi Qiu, Laurent Younes.
Description: New Jersey : World Scientific, [2020] | Series: Lecture notes series, Institute for
    Mathematical Sciences, National University of Singapore, 1793-0758 ; vol. 37 |
    "This volume results from the program "Mathematics of Shapes and Applications"
    that took place in Singapore, on 4–31 July, 2016 at the Institute for Mathematical Sciences,
    National University of Singapore (IMS NUS)"--Preface. | Includes bibliographical references.
Identifiers: LCCN 2019030201 | ISBN 9789811200120 (hardcover)
Subjects: LCSH: Shape theory (Topology) | Image analysis. | Image processing--Digital techniques.
Classification: LCC QA612.7 .M355 2020 | DDC 514/.24--dc23
LC record available at https://lccn.loc.gov/2019030201

British Library Cataloguing-in-Publication Data
A catalogue record for this book is available from the British Library.

For any available supplementary material, please visit
https://www.worldscientific.com/worldscibooks/10.1142/11272#t=suppl

Contents

Foreword

The Institute for Mathematical Sciences (IMS) organizes thematic programs of longer duration as well as shorter workshops and public lectures. The themes are selected from among areas at the forefront of current research in the mathematical sciences and their applications. Each volume of the *IMS Lecture Notes Series* is a compendium of papers based on lectures and tutorials delivered at the IMS. The aim is to make original papers and expository articles on a subject of current interest available to the international research community. These volumes also serve as a record of activities that took place at the IMS. We hope that through regular publication of these Lecture Notes the Institute will achieve, in part, its objective of reaching out to the community of scholars in the promotion of research in the mathematical sciences.

Chitat Chong
Adrian Röllin
Series Editors

Preface

This volume results from the program "Mathematics of Shapes and Applications" that took place in Singapore, on 4–31 July, 2016 at the Institute for Mathematical Sciences, National University of Singapore (IMS NUS). The program consisted of two weeks of summer schools on the mathematics of shapes and two week-long workshop sessions titled "State-of-the-Art Shape Research and its Applications" and "Applications: Biomedical Imaging and Computer Vision." The two workshops included 46 presentations, and attracted more than 100 local and overseas participants: PhD students, postdocs, researchers and professors.

Not forgetting important precursors, such as D'Arcy Thompson and his seminal treatise "On Growth and Forms," mathematical formulations of shape spaces can be traced back to the works of David Kendall on shape manifolds formed by cosets of configurations of points, and of Ulf Grenander on deformable templates, both occurring in the 1980s. Such works rapidly found application areas, in spatial statistics, computer vision and medical imaging and were quickly followed by new developments in a constant interaction between mathematics and applications. This led to new contributions to the study of Riemannian manifolds of curves or surfaces, or of diffeomorphism groups, to the construction of new and powerful shape or image registration algorithms, of new techniques for the statistical analysis of datasets of shapes, and to the emergence of new applied disciplines, such as computational anatomy, pioneered by Ulf Grenander and Michael Miller, which was an important component of the second part of the program.

This volume comprises eight novel papers that were presented at the workshop sessions, or resulting from the stimulating discussions occurring during the workshops and tutorials. These papers cover various active research areas in shape analysis and medical imaging, including fundamental results on shape theory, new methods for shape registration, recent developments on stochastic shape evolution, or improvements to the analysis and representation of medical images. We would like to thank all the participants for contributing to making the program a lively one, and we would like to thank all the reviewers of the papers for the present volume.

The editors would like to express gratitude to the IMS NUS, its staff, its advisory board for funding, organizing and providing an excellent support that made the "Mathematics of Shapes and Applications" program possible.

Organizing Committee:

> Co-chairs: Hui Ji (National University of Singapore), Sergey Kushnarev (Singapore University of Technology and Design), Laurent Younes (Johns Hopkins University).
> Members: Alvina Goh (National University of Singapore), Anqi Qiu (National University of Singapore).

> Sergey Kushnarev, Anqi Qiu and Laurent Younes
> *Volume Editors*

Variational Second-Order Interpolation on the Group of Diffeomorphisms with a Right-Invariant Metric

François-Xavier Vialard

Université Paris-Est, LIGM (UMR 8049),
UPEM, F77454 Marne-la-Vallée, France
francois-xavier.vialard@u-pem.fr

In this note, we propose a variational framework in which the minimization of the acceleration on the group of diffeomorphisms endowed with a right-invariant metric is well-posed. It relies on constraining the acceleration to belong to a Sobolev space of order higher than the order of the metric in order to gain compactness. It provides the theoretical guarantee of existence of minimizers which is compulsory for numerical simulations.

1. Introduction

The question of interpolating a time-sequence of shapes with a smooth curve in shape space and representing a shape evolution with few parameters have been addressed in the literature related to shape analysis and medical imaging since the last ten years. Several methods have been proposed and studied and they essentially rely on extensions of standard tools available in Euclidean geometry to shape spaces. In this direction, we mention geodesic regression, cubic regression, kernel methods... The generalization of these tools to infinite dimensional setting are sometimes complicated by the fact that the shape space is not a flat space, nor a finite dimensional space. However, the shape space is usually endowed with a Riemannian structure and most often in infinite dimensions. The generalizations of these Euclidean tools are often introduced by variational formulations, the simplest example being the case of shortest path between two shapes, i.e. geodesics on the space of shapes. Even in that particular example, finding a variational/analytical setting in which minimizing geodesics are well defined is of interest, since the existence of an extremum is not guaranteed in general and is complicated by the infinite dimensional setting. For ex-

1

ample, in the case of group of diffeomorphisms, this question is addressed in [5], in which the authors prove that the group of diffeomorphisms of the Euclidean space endowed with a right-invariant Sobolev metric of high enough order is complete in the sense of the Hopf-Rinow theorem. The case of the group of diffeomorphisms with right-invariant metric is relevant for applications in medical imaging and in particular for the problem of diffeomorphic image matching [26, 23, 2]. It is also natural to study and develop higher-order interpolations in the space of shapes, which has been actively developed in finite dimensions [3, 16, 19, 6, 9, 13, 15]. It was also extensively used and numerically developed in image processing and computer vision [18, 7, 24, 20, 4, 8]. In the past few years, these higher-order models have been introduced in biomedical imaging for interpolation of a time sequence of shapes. They have been proposed in [25] for a diffeomorphic group action on a finite dimensional manifold and further developed for general invariant higher-order Lagrangians in [11], [12] on a group. A numerical implementation together with a generalized model have been proposed in [21] in the context of medical imaging applications. However, in all these articles, the question of existence of an extremum is not treated. An attempt is given in [22] where the exact relaxation of the problem is shown on the group of diffeomorphisms of the interval $[0, 1]$. The main result of [22] consists in providing the existence of a minimizer in a larger space where the relaxation is defined. Although it does not completely solve the problem, it shows that existence of cubic splines for a group of diffeomorphisms with a right-invariant metric is nontrivial. Let us discuss where the difficulty comes from in a Riemannian setting. Riemannian splines are minimizers of

$$\mathcal{J}(x) = \int_0^1 g\left(\frac{D}{Dt}\dot{x}, \frac{D}{Dt}\dot{x}\right) \, dt, \qquad (1.1)$$

where (M, g) is a Riemannian manifold, $\frac{D}{Dt}$ is its associated covariant derivative and x is a sufficiently smooth curve from $[0, 1]$ in M satisfying first-order boundary conditions, i.e. $x(0), \dot{x}(0)$ and $x(1), \dot{x}(1)$ are fixed. The term

$$\frac{D}{Dt}\dot{x} = \ddot{x} + \Gamma(x)(\dot{x}, \dot{x})$$

(written in coordinates, with Γ the Christoffel symbols) contains nonlinearities which contribute in the variational problem (1.1) to possibly generating high-frequency oscillations in the space variable.

Although this notion of Riemannian cubics could not be well defined in general, it is possible to slightly modify it to make it well-posed. A modification of this type has recently been proposed in [14] in their framework. In this paper, we propose a simple variational setting which makes the second-order variational problem well-posed at the expense of increasing the regularity of the group of diffeomorphisms on which the second-order interpolation is feasible. For practical applications, this gain of smoothness, or loss of controllability of the diffeomorphism does not matter so much since smoothness is preferred in medical image registration context. However, the theoretical existence is guaranteed.

The main result of the paper is the following

Theorem 1.1: *Let Ω be a bounded domain in $\mathbb{R}^d$ and $s > d/2 + 1$ and $s' \geq s + 1$ be positive real numbers. Let $\mathcal{G}_{H^{s'}(\Omega,\mathbb{R}^d)}$ be the group of Sobolev diffeomorphism of order s' and $(x_0, v_0), (x_1, v_1) \in T\mathcal{G}_{H^{s'}(\Omega,\mathbb{R}^d)}$ two elements in the tangent bundle of the group.*

There exists a minimizer to the functional

$$\mathcal{J}(x) = \int_0^1 \left\| \frac{D}{Dt} \dot{x} \circ x^{-1} \right\|_{H^{s'}}^2 \, dt \,, \tag{1.2}$$

on the space of curves $x(t) \in \mathcal{G}_{H^{s'}(\Omega,\mathbb{R}^d)}$ that are H^2 in time and that satisfy the first-order time boundary conditions $(x(0), \dot{x}(0)) = (x_0, v_0)$ and $(x(1), \dot{x}(1)) = (x_1, v_1)$. In addition, $\frac{D}{Dt}$ denotes the covariant derivative associated with the right-invariant H^s metric.

The notation $\mathcal{G}_{H^{s'}(\Omega,\mathbb{R}^d)}$ stands for the diffeomorphism group of Sobolev diffeomorphisms of order s', as explained in Section 2. In other words, the acceleration is penalized with a stronger metric than the one which is used to compute it. The rest of the paper is devoted to prove this theorem. In a nutshell, it is based on the idea that the stronger norm prevents the minimizing sequence from creating oscillations, e.g. it gives enough compactness.

2. Background on right-invariant metrics on diffeomorphisms group

Sobolev right-invariant metrics on the group of diffeomorphisms. We present hereafter the main result of [5]. Let M be either $\mathbb{R}^d$ or a compact manifold without boundary of dimension d. We define hereafter a group of diffeomorphisms of M which is a complete metric space. Consider a space V a Hilbert space of vector fields on M (rapidly decreasing at infinity in the unbounded case), which are tangent to the boundary of M if there is any, such that the inclusion map $V \hookrightarrow W^{1,\infty}(M, \mathbb{R}^d)$ is continuous. This hypothesis implies that the flow of a time dependent vector field in $L^2([0, 1], V)$ is well defined, see [26] , and it is at every time a diffeomorphism of M. Then, the set of flows at time 1 defines a group of diffeomorphisms denoted by $\mathcal{G}_V$. Denoting

$$\mathrm{Fl}_t(\xi) = \varphi(t) \tag{2.1}$$

where φ solves the flow equation

$$\partial_t \varphi(t, x) = \xi(t, \varphi(t, x)) \tag{2.2}$$
$$\varphi(0, x) = x \;\; \forall x \in D \,, \tag{2.3}$$

we define

$$\mathcal{G}_V \stackrel{\mathrm{def.}}{=} \{\varphi(1) \,:\, \exists \xi \in L^2([0,1], V) \text{ s.t. } \mathrm{Fl}_1(\xi)\} \,, \tag{2.4}$$

which has been introduced by Trouvé in [23]. On this group, Trouvé defines a metric

$$\mathrm{dist}(\psi_1, \psi_0)^2 = \inf \left\{ \int_0^1 \|\xi\|_V^2 \, dt \,:\, \xi \in L^2([0,1], V) \text{ s.t. } \psi_1 = \mathrm{Fl}_1(\xi) \circ \psi_0 \right\} \tag{2.5}$$

under which he proves that $\mathcal{G}_V$ is complete. This construction is also valid when M is a bounded domain in $\mathbb{R}^d$. In full generality, that is for a general space of vector fields V, very few properties are known on this group. For instance, it is *a priori* not a topological group, or more precisely, there is no known topological structure making it a topological group (the inversion need not be continuous). Moreover, there does not need to be a differentiable structure on this group. However, for certain choices of spaces V, such structures are available and therefore more properties can be derived in this situation. Indeed, consider the group $\mathcal{D}^s(M)$, with $s > d/2 + 1$, which consists of all C^1-diffeomorphisms of Sobolev regularity H^s. It is known since the work of Ebin and Marsden [10] that $\mathcal{D}^s(M)$ is a smooth Hilbert manifold and a topological group. The main result in [5] shows that $\mathcal{G}_{H^s} = \mathcal{D}^s(M)_0$, where the subscript stands for the connected component of identity.

Theorem 2.1: *Let M be $\mathbb{R}^d$ or a closed manifold and $s > d/2 + 1$. If G^s is a smooth, right-invariant Sobolev-metric of order s on $\mathcal{D}^s(M)$, then*

(1) $(\mathcal{D}^s(M), G^s)$ is geodesically complete;
(2) $(\mathcal{D}^s(M)_0, \mathrm{dist}^s)$ is a complete metric space;
(3) Any two elements of $\mathcal{D}^s(M)_0$ can be joined by a minimizing geodesic.

The statements also hold for the subgroups $\mathcal{D}_\mu^s(M)$ and $\mathcal{D}_\omega^s(M)$ of diffeomorphisms preserving a volume form μ or a symplectic structure ω.

The crucial ingredient in the proof is showing that the flow map

$$\mathrm{Fl}_t : L^1(I, \mathfrak{X}^s(M)) \to \mathcal{D}^s(M) \tag{2.6}$$

exists and is continuous.

In [25], we introduced the use of cubic splines *in the space of shapes* to interpolate a sequence of shapes that are time dependent. Riemannian cubics (also called Riemannian splines) and probably more famous, its constrained alternative called Elastica belong to a class of problems that have been studied since the work of Euler (see the discussion in [18]). Let us present the variational problem in a Riemannian setting. Riemannian splines are minimizers of

$$\mathcal{J}(x) = \int_0^1 g\left(\frac{D}{Dt}\dot{x}, \frac{D}{Dt}\dot{x}\right) \, dt \,, \tag{2.7}$$

where (M, g) is a Riemannian manifold, $\frac{D}{Dt}$ is its associated covariant derivative and x is a sufficiently smooth curve from $[0,1]$ in M satisfying first-order boundary conditions, i.e. $x(0), \dot{x}(0)$ and $x(1), \dot{x}(1)$ are fixed. The case of Elastica consists in restricting the previous optimization problem to the set of curves that are parametrized by unit speed (when the problem is feasible), namely $g(\dot{x}, \dot{x}) = 1$ for all time. To the best of our knowledge, the only paper that deals with analytical questions is [13] where the authors show in particular the existence of minimizers of a second-order functional on the space curves on a complete finite dimensional Riemannian manifold.

In [11], higher-order models are proposed on groups of diffeomorphisms but for the standard Riemannian cubics functional, no analytical study was provided. Indeed, in the case of a Lie group G (and $\mathfrak{g}$ its Lie algebra) with a right-invariant metric ($\|\cdot\|_\mathfrak{g}$ denoting the norm on the Lie algebra), the covariant derivative can be written as follows: Let $V(t) \in T_{g(t)}G$ be a vector field along a curve $g(t) \in G$

$$\frac{D}{Dt}V = \left(\dot{\nu} + \frac{1}{2}\mathrm{ad}_\xi^\dagger \nu + \frac{1}{2}\mathrm{ad}_\nu^\dagger \xi - \frac{1}{2}[\xi, \nu]\right)_G(g)\,, \tag{2.8}$$

where $\mathrm{ad}^\dagger$ is the metric adjoint defined by

$$\mathrm{ad}_\nu^\dagger \kappa := (\mathrm{ad}_\nu^*(\kappa^\flat))^\sharp \tag{2.9}$$

for any $\nu, \kappa \in \mathfrak{g}$ and $\flat$ and $\sharp$ are the musical operator for the cometric and metric operator[a]. Therefore, the reduced lagrangian for (2.7) is

$$\mathcal{J}(x) = \int_0^1 \|\dot{\xi} + \mathrm{ad}_\xi^\dagger \xi\|_{\mathfrak{g}}^2 \, dt \,, \tag{2.10}$$

where $\mathrm{ad}^\dagger$ is the metric adjoint, i.e., it is written as

$$\mathrm{ad}_\nu^\dagger \kappa := (\mathrm{ad}_\nu^*(\kappa^\flat))^\sharp \tag{2.11}$$

for any $\nu, \kappa \in \mathfrak{g}$. We can also formulate the variational problem on the dual of the Lie algebra $\mathfrak{g}^*$ by

$$\mathcal{J}(x) = \int_0^1 \|a(t)\|_{\mathfrak{g}^*}^2 \, dt \,, \tag{2.12}$$

under the constraint

$$\dot{m} + \mathrm{ad}_\xi^* m = a \,. \tag{2.13}$$

In infinite dimensions, there is a clear obstacle to use reduction since the operator $\mathrm{ad}^\dagger$ is unbounded on the tangent space at identity due to a loss of derivative. However, using the smooth Riemannian structure on $\mathcal{D}^s$, functional (2.7) is well defined.

The following proposition of [13] is valid in infinite dimensions:

Proposition 2.2: *Let (M, g) be an infinite dimensional Riemannian manifold and*

$$\Omega_{0,1}(M) := \{x \in H^2([0,1], M) \,|\, x(i) = x_i \,, \, \dot{x}(i) = v_i \text{ for } i = 0, 1\}$$

be the space of paths with the first-order boundary constraints for given $(x_0, v_0) \in TM$ and $(x_1, v_1) \in TM$. The functional (2.7) is smooth on $\Omega_{0,1}(M)$ and

$$\mathcal{J}'(x)(v) = \int_0^1 g\left(\frac{D^2}{Dt^2}\dot{v}, \frac{D}{Dt}\dot{x}\right) - g\left(R\left(\dot{x}, \frac{D}{Dt}\dot{x}\right), v\right) \, dt \,. \tag{2.14}$$

A critical point of $\mathcal{J}$ is a smooth curve that satisfies the Riemannian cubic equation

$$\frac{D^3}{Dt^3}\dot{x} - R\left(\dot{x}, \frac{D}{Dt}\dot{x}\right)\dot{x} = 0 \,. \tag{2.15}$$

The critical points of $\mathcal{J}$ are sometimes called Riemannian cubics or cubic polynomials. The existence of minimizers does not follow from the corresponding proof in [13] since it strongly relies on the finite dimension hypothesis to have compactness properties.

---

[a]The $\sharp$ operator is the inverse of the $\flat$ operator which is a map from the tangent to the cotangent space of M, $\flat : TM \mapsto T^*M$ given in coordinates by $v^\flat \stackrel{\text{def.}}{=} g(v, \cdot) \in T^*M$.

3. The main result

We formulate the main result on the flat torus but it can be generalized to bounded domains in $\mathbb{R}^d$ in a straightforward way.

Theorem 3.1: *Let* $\mathbb{T}^d \overset{\text{def}}{=} \mathbb{R}^d/\mathbb{Z}^d$ *be the* d *dimensional flat torus and* $s' \geq s + 1$. *There exists a minimizer to the functional*

$$\mathcal{J}(x) = \int_0^1 \left\| \frac{D}{Dt}\dot\varphi \circ \varphi^{-1} \right\|_{H^{s'}}^2 \, dt \,, \tag{3.1}$$

on the loop space $\Omega_{0,1}(\mathcal{G}_{H^{s'}}(\mathbb{T}^d,\mathbb{R}^d))$ *where* $\frac{D}{Dt}$ *is the covariant derivative associated with the right-invariant* H^s *metric.*

Before proving the theorem, we prove the following lemmas:

Lemma 3.2: *Let* $\alpha \in L^2([0,1], H^s)$, *then there exists a unique solution, defined for* $t \in [0,1]$, *to the system*

$$\dot\varphi = v \tag{3.2a}$$
$$\dot v = -\Gamma(\varphi)(v,v) + \alpha \circ \varphi \,, \tag{3.2b}$$

for given initial conditions $\varphi(0) = \varphi_0 \in D^{s+1}$ *and* $v(0) = v_0 \in H^{s+1}$.

Proof: Solutions exist for short time since the system is Lipschitz on $\mathcal{D}^s \times H^s$ and the existence theorem for Caratheodory equation gives the result. Existence for all time is not guaranteed *a priori* since the second equation is quadratic in v. Let us denote $u = v \circ \varphi^{-1}$ and $f(t) := \frac{1}{2}g(\varphi)(v,v) = \frac{1}{2}\|u\|_{H^s}^2$. Taking derivative of f gives $f'(t) = \langle \alpha(t), u(t) \rangle_{H^s}$, so that by the Cauchy-Schwarz inequality, we get

$$f(t) \leq \int_0^t \sqrt{f(z)} \|\alpha\|_{H^s} \, dz \leq \|\alpha\|_{L^2([0,1],H^s)} \sqrt{\int_0^t f(z)\,dz} \tag{3.3}$$

$$f(t) \leq \|\alpha\|_{L^2([0,1],H^s)} \left(1 + \int_0^t f(z)\,dz \right). \tag{3.4}$$

Using Gronwall's lemma, it implies that $f(t)$ is bounded for on $[0,T]$ where T is the supremum (possible blow-up) time of definition. Therefore, $r := \int_0^T \|u\|_{H^s}^2 \, dt = \int_0^T f(t)\,dt < \infty$ which means that $u \in L^2([0,T], H^s)$. As a consequence, for all time $t \in [0,T[$, $\varphi(t) \in B(x_0, r)$, which is the ball of radius r for the geodesic distance on D^s. In addition, $\lim_{t \to T} \varphi(t)$ is well defined since D^s is metrically complete. Remark that $\Gamma(\varphi(t))$ is bounded (uniformly in time) as an operator on $H^s \times H^s$ since Γ is smooth on $\mathcal{D}^s$

and thus continuous on the path $\varphi(t)$. In particular, the right-hand side of the equation (3.2b) belongs to $L^1([0,T], H^s)$ so that

$$v(T) := \int_0^T \left(-\Gamma(\varphi(t))(v(t), v(t)) + \alpha(t) \circ \varphi(t) \right) dt. \qquad (3.5)$$

Using short time existence on $(\varphi(T), v(T))$, the solution can be extended for short time from time T so that in fact $T = 1$. $\qquad\square$

In the following lemma, we study the solutions of the system (3.12) but written on the dual of the tangent space at identity. As mentioned in Section 2, one can rewrite the minimization as in Equation (2.12), however the "dual" acceleration is measured using with the corresponding dual norm. In our case, the dual norm (w.r.t. to H^s) associated with $H^{s+1} \subset H^s$ is $(H^{s-1})^* \subset (H^s)^*$ as can be seen by a direct computation in Fourier spaces.

Lemma 3.3: *Let $a \in L^2([0,1], (H^{s-2})^*)$ and K_s the kernel associated with the H^s norm. Then, the following integral equation, with initial condition $g(0) = \mathrm{Id}$ and $m(0) \in (H^{s-2})^*$,*

$$\begin{cases} \dot{g}(t) = (K_s m(t)) \circ g(t) \\ m(t) = Ad^*_{g(t)^{-1}}(m(0)) + \int_0^t Ad^*_{g(s) \circ g(t)^{-1}}(a(s)) ds, \end{cases} \qquad (3.6)$$

has a unique solution in $m(t) \in C^0([0,T], (H^s)^*)$. If $a \in L^2([0,1], (H^{s-1})^*)$, then there exists a solution to the integral equation with initial condition in $m(0) \in (H^{s-1})^*$.

Proof: The proof of the existence for $a \in L^2([0,1], (H^{s-2})^*)$ follows a standard fixed point method. Let $\Psi : C^0([0,T], (H^s)^*) \to C^0([0,T], (H^s)^*)$ be defined by Formula (3.6), namely

$$\Psi(m)(t) = Ad^*_{g(t)^{-1}}(m(0)) + \int_0^t Ad^*_{g(s) \circ g(t)^{-1}}(a(s)) ds. \qquad (3.7)$$

Remark first that $\Psi(m)$ belongs to $C^0([0,T], (H^s)^*)$. Moreover, the right-hand side of Equation (3.7) is well-defined since the integrand is integrable. Now we claim that this map is a contraction on $(C^0([0,T], (H^s)^*), \|\cdot\|_\infty)$ for T small enough,

$$\|\Psi(m_1)(t) - \Psi(m_2)(t)\|_\infty \leq \sup_{t \in [0,T]} \left(\|Ad^*_{g_1(t)^{-1}}(m(0)) - Ad^*_{g_2(t)^{-1}}(m(0))\|_{(H^s)^*} \right.$$

$$\left. + \int_0^t \|Ad^*_{g_1(s) \circ g_1(t)^{-1}} a(s) - Ad^*_{g_2(s) \circ g_2(t)^{-1}} a(s)\|_{(H^s)^*} ds \right). \qquad (3.8)$$

We need to estimate for $\alpha \in (H^{s-2})^*$ and $w \in H^s$. There exists a constant $M_0 > 0$ such that

$$\langle Ad^*_{g_1}(\alpha) - Ad^*_{g_2}(\alpha), w \rangle_{L^2} \leq \|\alpha\|_{(H^{s-2})^*} \|Ad_{g_1}(w) - Ad_{g_2}(w)\|_{H^{s-2}}$$
$$M_0 \|\alpha\|_{(H^{s-2})^*} \|g_1 - g_2\|_{H^{s-2}} \|w\|_{H^s}.$$

Moreover, there exists a constant $M_1 > 0$ s.t.

$$\sup_{s,t \in [0,T]} \|g_1(s) \circ g_1(t)^{-1} - g_2(s) \circ g_2(t)^{-1}\|_{H^{s-2}} \leq M_1 \|m_1 - m_2\|_{L^2([0,T],(H^s)^*)},$$

which follows from [5]. Therefore, there exists a constant $M_2 > 0$ s.t.

$$\|\Psi(m_1) - \Psi(m_2)(t)\|_{\infty} \leq M_2 \sqrt{t} \|m_1 - m_2\|_{\infty} \left(\|m(0)\|_{(H^s)^*} + \int_0^T \|a(s)\|_{(H^s)^*} \, ds \right). \tag{3.9}$$

Hence the map Ψ is a contraction for t small enough and it therefore proves the existence and uniqueness of a solution of formula (3.6) for short times. Remark that Equation (3.9) gives an upper bound $t_{lip} := \left(M_2(\|m(0)\|_{W_2^*} + \int_0^T \|a(s)\|_{W_2^*} \, ds) \right)^{-2}$ such that for every $t < t_{lip}$, Ψ is a contraction. In addition this upper bound is valid at any time $t \in [0, T]$. Then the existence and uniqueness until time T follows straightforwardly by an iterative application of the short-time result.

The second part of the proof consists in showing existence of solutions for $a \in L^2([0,1], (H^{s-1})^*)$ which is done using a compactness argument. Let a_n converging to a in $L^2([0,1], (H^{s-1})^*)$, then the solution $m_n \in C^0([0,T], (H^s)^*)$ actually belongs to $H^1([0,1], (H^{s-1})^*)$ because $\|Ad^*_g(m)\|_{(H^{s-1})^*} \leq M_2 \|g\|_{H^s} \|m\|_{(H^{s-1})^*}$. By the Aubin-Lions-Simon theorem, see introduction of [17] for its statement, $H^1([0,1], (H^{s-1})^*)$ is compactly embedded in $L^2([0,1], (H^s)^*)$, thus one can extract a strongly convergent sequence in $L^2([0,1], (H^s)^*)$. By Theorem 2.1, the flow associated with the momentum m_n, denoted by $g^n_{s,t}$ strongly converges in H^s (actually uniformly in s, t) to $g_{s,t}$ the flow associated with the limit m. Then, it implies that the integrand $Ad^*_{g^n_{t,s}}(a_n(s))$ in Formula (3.7) converges to $Ad^*_{g_{t,s}}(a(s))$ in $(H^{s-1})^*$. Since the integrand is bounded uniformly, the Lebesgue convergence theorem applies and the result is obtained. $\square$

Remark 3.4:

(1) The reason why we are not able to treat the case of $a \in L^2([0,1], (H^{s-1})^*)$ is because the flow map in Theorem 2.1 is only continuous and (possibly) not Lipschitz.

(2) Note that Lemma 3.3 can be considered as the Eulerian version of
Lemma 3.2 and the latter achieves a better result since uniqueness of
the solution is proven in H^{s+1}. However, our proof of the main theorem
will require the use of Lemma 3.3 which gives the fact that the Eulerian
velocity of the solutions of Lemma 3.2 are bounded in $H^1([0,1], (H^s)^*)$.

Lemma 3.5: *Let $s > d/2 + 1$, $\alpha_n \in H^s$ weakly converging to α and $\varphi_n \in$
Diff^s which strongly converges to φ, then the composition $\alpha_n \circ \varphi_n$ weakly
converges to $\alpha \circ \varphi$.*

Proof: We prove the weak convergence by proving that the sequence is
bounded and that it weakly converges on a dense set of H^s. First, remark
that $\|\alpha_n \circ \varphi_n\|_{H^s}$ is bounded in H^s since the composition by a diffeomor-
phism in Diff^s is bounded (see [5]).

Let $m \in (H^s)^* \cap \mathcal{M}$ where $\mathcal{M}$ denotes the space of Radon measures,
consider $\langle \alpha_n \circ \varphi_n, m \rangle_{L^2}$, which can be written by a change of variable as

$$\langle \alpha_n \circ \varphi_n, m \rangle_{L^2} = \langle \alpha_n, (\varphi_n)_*(m) \rangle_{L^2}, \qquad (3.10)$$

where $(\varphi_n)_*(m)$ is the pushforward of m by φ_n. Since φ_n is strongly convergent in Diff^s, we have that $(\varphi_n)_*(m)$ strongly converge in $(H^s)^*$ to $\varphi_*(m)$.
Therefore, $\langle \alpha_n, (\varphi_n)_*(m) \rangle_{L^2}$ converges to $\langle \alpha, \varphi_*(m) \rangle_{L^2}$, which gives the
result.   □

Proof: [Proof of the theorem] First note that the space $\Omega_{0,1}(\mathcal{G}_{H^{s'}(\Omega,\mathbb{R}^d)})$
is non-empty, i.e. there exists a path in the group which is H^2 in time and
which satisfies the constraints at times 0 and 1. Let us consider a geodesic
path $c(t)$ between $\varphi(0)$ and $\varphi(1)$, by concatenation of paths with boundary conditions on positions and velocities, it is then sufficient to connect
$(\varphi(0), \dot{\varphi}(0))$ and $(c(0), \dot{c}(0))$ with a smooth curve. Using again concatenation of paths and right-invariance, it is sufficient to connect, for every
$v_0 \in T_{\mathrm{Id}}\mathcal{G}_{H^{s'}(\Omega,\mathbb{R}^d)}$, $(\mathrm{Id}, 0)$ and (Id, v_0) with a smooth path. It can be done
using time reparametrization of the geodesic $\psi(t)$ starting from Id with
initial velocity v_0. This geodesic is defined for all times $t \in (-\infty, +\infty)$
by geodesic completeness of the metric. The chosen interpolating curve is
now given by $s(t) \stackrel{\text{def.}}{=} \psi(t^3 - t^2)$. Note that the cubic polynomial function
$t \mapsto t^3 - t^2$ takes negative values which is allowed by the completeness
property. By direct computation, it can be checked that $s(0) = \mathrm{Id}$, $\dot{s}(0) = 0$
and $s(1) = \mathrm{Id}$, $\dot{s}(1) = \dot{\psi}(0) = v_0$.

We will use natural coordinates, i.e. $\mathrm{Id} + H^s$ to describe elements of the
loop space $\Omega_{0,1}(\mathcal{G}_{H^{s'}(\Omega,\mathbb{R}^d)})$. The term $\left\|\frac{D}{Dt}\dot{\varphi} \circ \varphi^{-1}\right\|_{H^{s'}}^2$ can be written in

coordinates

$$\| (\ddot{\varphi} + \Gamma(\varphi)(\dot{\varphi}, \dot{\varphi})) \circ \varphi^{-1} \|_{H^{s'}}^2$$

so that, it is natural to introduce the change of variable $(\dot{v} + \Gamma(\varphi)(v, v)) \circ \varphi^{-1} := \alpha$ where $\dot{\varphi} = v$. The variational problem (2.7) can be rewritten as the minimization of the functional defined on the Hilbert space $L^2([0, 1], H^{s'})$

$$\ell(\alpha) = \int_0^1 \|\alpha(t)\|_{H^{s'}}^2 \, dt, \tag{3.11}$$

under the constraint

$$\begin{cases} \dot{\varphi} = v \\ \dot{v} = -\Gamma(\varphi)(v, v) + \alpha \circ \varphi, \end{cases} \tag{3.12}$$

and the boundary conditions, $\varphi(0) = \varphi_0$, $\varphi(1) = \varphi_1$ and $v(0) = v_0$, $v(1) = v_1$.

This functional is lower semi-continuous on $L^2([0, 1], H^{s'})$. Let α_n be a minimizing sequence in $L^2([0, 1], H^{s'})$ weakly converging to α. The condition to be checked is that the weak limit satisfies the constraints. Using Lemma 3.3, the sequence $v_n \circ \varphi_n^{-1} \in H^1([0, 1], H^{s'})$ is bounded. Since $H^1([0, 1], H^{s'})$ is compactly embedded in $C^0([0, 1], H^s)$, one can extract a strongly converging sequence in $C^0([0, 1], H^s)$ which implies the strong convergence of $\varphi_n(1)$ and $v_n(1)$ in H^s. The first consequence is that the boundary constraints $\varphi(1), v(1)$ are satisfied at the limit. It also implies that the term $\Gamma(\varphi_n)(v_n, v_n)$ is strongly convergent in H^s and the term $\alpha_n \circ \varphi_n$ is weakly convergent to $\alpha \circ \varphi$, using Lemma 3.5. Therefore, we have the equality $v(t) = v(0) + \int_0^t (-\Gamma(\varphi(s))(v(s), v(s)) + \alpha \circ \varphi) \, ds$, which implies that the couple (φ, v) is the solution of the integral equation associated with System (3.12).   □

We give hereafter a theorem on spline interpolation of time sequences.

Theorem 3.6: *Let $\varphi_1, \ldots, \varphi_n$ be n diffeomorphisms in Diff_0^{s+1} and $t_1 < \cdots < t_n$ be a sequence of n positive reals. There exists a path $\varphi(t) \in \mathrm{Diff}_0^{s+1}$, which minimize the acceleration functional*

$$\|\dot{\varphi}(0) \circ \varphi^{-1}\|_{H^{s'}}^2 + \int_{t_1}^{t_n} \left\| \frac{D}{Dt} \dot{\varphi} \circ \varphi^{-1} \right\|_{H^{s'}}^2 \, dt, \tag{3.13}$$

among all curves satisfying $\varphi(t_i) = \varphi_i$ for $i \in 1, \ldots, n$.

Proof: The proof is similar to that of Theorem 3.1 and we do not repeat the arguments here. Note that the penalization on the initial speed seems necessary in order for the curve to stay in a bounded metric ball.[b]   □

4. Conclusion

This proof of existence was provided to fill in the gap between theory and application of the variational models proposed in [21]. From a theoretical point of view, we leave open the question if this approach could be adapted for $s' > s$ rather than the more demanding condition $s' \geq s + 1$.

In addition, we have not treated the case of the Riemannian metric induced by the action of the diffeomorphism group on the space of images as usually used in the LDDMM setting [2], which was also implemented in [21]. However, with minor modifications, our approach can possibly be adapted.

Last, note that the point of view taken here could be generalized to other infinite dimensional shape spaces where, very often, the available Riemannian metrics are defined by some degree of smoothing of tangent vectors, e.g. Sobolev intrinsic metrics on the space of curves or surfaces [1].

References

1. M. Bauer, P. Harms, and P. W. Michor. Sobolev metrics on shape space, ii: Weighted sobolev metrics and almost local metrics. *J. Geom. Mech.*, 4(4):365–383, 2012.
2. M. Faisal Beg, Michael I. Miller, Alain Trouvé, and Laurent Younes. Computing large deformation metric mappings via geodesic flow of diffeomorphisms. *International Journal of Computer Vision*, 61:139–157, 2005.
3. G. Birkhoff, H. Burchard, and D. Thomas. Non-linear interpolation by splines, pseudosplines and elastica. *General Motors Research Lab report*, 468, 1965.
4. Guido Brunnett and Johannes Kiefer. Interpolation with minimal-energy splines. *Computer-Aided Design*, 26(2):137–144, 1994.
5. Martins Bruveris and François-Xavier Vialard. On completeness of groups of diffeomorphisms. *J. Eur. Math. Soc. (JEMS)*, 19(5):1507–1544, 2017.
6. M. Camarinha, F. Silva Leite, and P.Crouch. Splines of class $\mathcal{C}^k$ on non-euclidean spaces. *IMA Journal of Mathematical Control & Information*, 12:399–410, 1995.

---

[b]On the flat 2D torus, straight lines with irrational slopes are dense and they can be parametrized with arbitrarily high velocity so that the infimum of (3.13) is 0 without speed penalization.

7. Frédéric Cao, Yann Gousseau, Simon Masnou, and Patrick Pérez. Geometrically guided exemplar-based inpainting. *SIAM Journal on Imaging Sciences*, 4(4):1143–1179, 2011.

8. Tony F. Chan, Sung Ha Kang, Kang, and Jianhong Shen. Euler's elastica and curvature based inpaintings. *SIAM J. Appl. Math*, 63:564–592, 2002.

9. P. Crouch and F. Silva Leite. The dynamic interpolation problem: On Riemannian manifold, Lie groups and symmetric spaces. *Journal of dynamical & Control Systems*, 1:177–202, 1995.

10. David G. Ebin and Jerrold Marsden. Groups of diffeomorphisms and the motion of an incompressible fluid. *Ann. of Math. (2)*, 92:102–163, 1970.

11. F. Gay-Balmaz, D. D. Holm, D. M. Meier, T. S. Ratiu, and F.-X. Vialard. Invariant Higher-Order Variational Problems. *Communications in Mathematical Physics*, 309:413–458, January 2012.

12. F. Gay-Balmaz, D. D. Holm, D. M. Meier, T. S. Ratiu, and F.-X. Vialard. Invariant Higher-Order Variational Problems II. *Journal of NonLinear Science*, 22:553–597, August 2012.

13. Roberto Giambo and Fabio Giannoni. An analytical theory for Riemannian cubic polynomials. *IMA Journal of Mathematical Control & Information*, 19:445–460, 2002.

14. B. Heeren, M. Rumpf, and B. Wirth. Variational time discretization of Riemannian splines. *ArXiv e-prints*, November 2017.

15. Norihito Koiso. Elasticae in a Riemannian submanifold. *Osaka J. Math.*, 29(3):539–543, 1992.

16. E. H. Lee and G. E. Forsythe. Variational study of nonlinear spline curves. *SIAM Review*, 15(1):120–133, 1973.

17. A. Moussa. Some variants of the classical Aubin–Lions lemma. *Journal of Evolution Equations*, 16(1):65–93, Mar 2016.

18. David Mumford. Elastica and computer vision. In Chandrajit L. Bajaj, editor, *Algebraic Geometry and its Applications*, pages 491–506. Springer New York, 1994.

19. L. Noakes, G. Heinzinger, and B. Paden. Cubic splines on curved spaces. *IMA Journal of Mathematical Control & Information*, 6:465–473, 1989.

20. Chafik Samir, P.-A. Absil, Anuj Srivastava, and Eric Klassen. A gradient-descent method for curve fitting on riemannian manifolds. *Foundations of Computational Mathematics*, 12(1):49–73, 2012.

21. Nikhil Singh, François-Xavier Vialard, and Marc Niethammer. Splines for diffeomorphisms. *Medical Image Analysis*, 25(1):56–71, 2015.

22. R. Tahraoui and F.-X. Vialard. Riemannian cubics on the group of diffeomorphisms and the Fisher-Rao metric. *ArXiv e-prints*, June 2016.

23. Alain Trouvé. Action de groupe de dimension infinie et reconnaissance de formes. (infinite dimensional group action and pattern recognition). 1995.

24. J. Ulen, P. Strandmark, and F. Kahl. Shortest paths with higher-order regularization. *Pattern Analysis and Machine Intelligence, IEEE Transactions on*, 37(12):2588–2600, Dec 2015.

25. F.-X. Vialard and A. Trouvé. Shape Splines and Stochastic Shape Evolutions: A Second Order Point of View. *Quart. Appl. Math.*, 2012.
26. Laurent Younes. *Shapes and Diffeomorphisms*. Springer, 2008.

Riemannian Geometry for Shape Analysis and Computational Anatomy

Martins Bruveris

Onfido,
3 Finsbury Avenue,
London EC2M 2PA, UK
martins.bruveris@gmail.com

Shape analysis and computational anatomy both make use of sophisticated tools from infinite-dimensional differential manifolds and Riemannian geometry on spaces of functions. While comprehensive references for the mathematical foundations exist, it is sometimes difficult to gain an overview on how differential geometry and functional analysis interact in a given problem. This paper aims to provide a roadmap to the uninitiated to the world of infinite-dimensional Riemannian manifolds, spaces of mappings and Sobolev metrics: all tools used in computational anatomy and shape analysis.

Contents

Introduction

These lecture notes were written to supplement a course given at the summer school "Mathematics of Shapes" in Singapore in July 2016. The aim of the course was to show how the language of differential geometry can be used in shape analysis and computational anatomy. Of course four lectures are not enough to fully do justice to the subject. However four lectures are enough to give an introduction to infinite-dimensional differential geometry, to show how the infinite-dimensional world differs from the finite-dimensional one and to point the interested reader to more in depth references. Infinite-dimensional differential geometry is treated in detail in [17, 15, 16] and more information about manifolds of maps can be found in the articles [5, 12]. For shape analysis one may consult [26] and for computational anatomy [28].

1. Infinite-dimensional manifolds and functional analysis

The first lecture gave an introduction to manifolds in infinite dimensions assuming familiarity with finite-dimensional differential geometry. Then it discussed the main differences between finite and infinite dimensions: loss of local compactness and no existence theorems for ODEs in Fréchet spaces. Finally it discussed Omori's theorem and what it means that the diffeomorphism group of a compact manifold cannot be modelled as a smooth Banach Lie group.

1.1. *Infinite-dimensional manifolds*

A *smooth manifold* modelled on the topological vector space E is a Hausdorff topological space M together with a family of *charts* $(u_\alpha, U_\alpha)_{\alpha \in A}$, such that

(1) $U_\alpha \subseteq M$ are open sets, $\bigcup_{\alpha \in A} U_\alpha = M$;
(2) $u_\alpha : U_\alpha \to u_\alpha(U_\alpha) \subseteq E$ are homeomorphisms onto open sets $u_\alpha(U_\alpha)$;
(3) $u_\beta \circ u_\alpha^{-1} : u_\alpha(U_\alpha \cap U_\beta) \to u_\beta(U_\alpha \cap U_\beta)$ are C^∞-smooth.

In this definition it does not matter, whether E is finite or infinite-dimensional. In fact, if E is finite-dimensional, then $E = \mathbb{R}^n$ for some $n \in \mathbb{N}$ and we recover the definition of a finite-dimensional manifold.

1.2. *Choice of a modelling space*

There are several classes of infinite-dimensional vector spaces to choose from. With increasing generality our space E can be a

(1) Hilbert space;
(2) Banach space;
(3) Fréchet space;
(4) convenient locally convex vector space.

We will assume basic familiarity with Hilbert and Banach spaces. All topological vector spaces are assumed to be Hausdorff. A *Fréchet space* is a locally convex topological vector space X, whose topology can be induced by a complete, translation-invariant metric, i.e., a metric $d : X \times X \to \mathbb{R}$ such that $d(x + h, y + h) = d(x, y)$. Alternatively a Fréchet space can be characterized as a Hausdorff topological space, whose topology may be induced by a countable family of seminorms $\| \cdot \|_n$, i.e., finite intersections of the sets $\{y : \|x - y\|_n < \varepsilon\}$ with some x, n, ε form a basis of the topology, and the topology is complete with respect to this family.

We will in these lectures ignore for the most part convenient vector spaces; a detailed exposition of manifolds modelled on these spaces can be found in [16]; we mention only that convenient vector spaces are necessary to model spaces of compactly supported functions on noncompact manifolds.

Each of these classes is more restrictive than the next one. A *Hilbert space* is a vector space with an inner product[a] $\langle \cdot, \cdot \rangle$. The inner product

---

[a]We will ignore questions of completeness in this informal discussion.

induces a norm via

$$\|x\| = \sqrt{\langle x, x \rangle}\,.$$

If we are just given a norm $\|\cdot\|$, then we have a *Banach space*. A norm can be used to define a distance d via

$$d(x, y) = \|x - y\|\,.$$

If we have only a distance function, then the space is a *Fréchet space*. It is in general not possible to go in the other direction.

1.3. *The Hilbert sphere*

A first example of an infinite-dimensional manifold is the unit sphere in a Hilbert space. Let E be an infinite-dimensional Hilbert space. Then

$$S = \{x \in E \,:\, \|x\| = 1\}$$

is a smooth manifold. We can construct charts on S in the following way: For $x_0 \in S$, define the subspace $E_{x_0} = \{y \in E \,:\, \langle y, x_0 \rangle = 0\}$, which is isomorphic to E itself. The chart map is given by

$$u_{x_0} : x \mapsto x - \langle x, x_0 \rangle x_0\,,$$

and is defined between the sets $U_{x_0} = \{x \in S \,:\, \langle x, x_0 \rangle > 0\}$ and $u_{x_0}(U_{x_0}) = \{y \in E_{x_0} \,:\, \|y\| < 1\}$. Its inverse is

$$u_{x_0}^{-1} : y \mapsto y + \sqrt{1 - \|y\|^2}\, x_0\,.$$

We will omit the verification that chart changes are smooth maps.

Note that if E is infinite-dimensional, then the sphere is not compact. To see this choose an orthonormal sequence $(e_n)_{n \in \mathbb{N}}$. Then $e_n \in S$, but the sequence does not contain a convergent subsequence, because $\|e_n - e_m\| = \sqrt{2}$. Hence S cannot be compact.

1.4. *The manifold* $\mathrm{Imm}(S^1, \mathbb{R}^d)$

One of the simplest spaces of functions is the space of smooth, periodic, immersed curves,

$$\mathrm{Imm}(S^1, \mathbb{R}^d) = \{c \in C^\infty(S^1, \mathbb{R}^d) \,:\, c'(\theta) \neq 0,\ \forall \theta \in S^1\}\,.$$

The modelling space for the manifold is $C^\infty(S^1, \mathbb{R}^d)$, the space of smooth, periodic functions[b]. When we talk about manifolds, we have to specify,

---

[b] A good introduction to the space of smooth functions on the circle can be found in [10].

what topology we mean. On the space $C^\infty(S^1, \mathbb{R}^d)$ of smooth functions we consider the topology of uniform convergence in all derivatives, i.e.,

$$f_n \to f \Leftrightarrow \lim_{n \to \infty} \|f_n^{(k)} - f^{(k)}\|_\infty = 0 \quad \forall k \in \mathbb{N},$$

where $\|f\|_\infty = \sup_{\theta \in S^1} |f(\theta)|$. A basis of open sets is formed by sets of the form

$$M(f, \varepsilon, k) = \left\{ g \in C^\infty(S^1, \mathbb{R}^d) : \|g^{(k)} - f^{(k)}\|_\infty < \varepsilon \right\},$$

with $\varepsilon \in \mathbb{R}_{>0}$ and $k \in \mathbb{N}$.

Proposition 1.1: $C^\infty(S^1, \mathbb{R}^d)$ *is a reflexive, nuclear, separable Fréchet space.*

See Sect. 6.2 in [24] and Thm. 4.4.12 in [24] for proofs. Having defined a topology, how does the set of immersions sit inside the set of all smooth functions?

Lemma 1.2: $\mathrm{Imm}(S^1, \mathbb{R}^d)$ *is an open subset of* $C^\infty(S^1, \mathbb{R}^d)$.

Proof: Given $f \in \mathrm{Imm}(S^1, \mathbb{R}^d)$, let $\varepsilon = \inf_{\theta \in S^1} |f'(\theta)|$. We have $\varepsilon > 0$ since S^1 is compact. Now consider the neighborhood $M(f, \varepsilon/2, 1)$ defined above; for $g \in M(f, \varepsilon/2, 1)$ we can estimate

$$|g'(\theta)| \geq |f'(\theta)| - \|g' - f'\|_\infty \geq \frac{\varepsilon}{2} > 0,$$

and thus $f \in M(f, \varepsilon/2, 1) \subseteq \mathrm{Imm}(S^1, \mathbb{R}^d)$. $\qquad \square$

See Thm. 2.1.1 in [13] for a more general statement about the spaces $\mathrm{Imm}(M, N)$. One can also show[c] that $\mathrm{Imm}(S^1, \mathbb{R}^d)$ is dense in $C^\infty(S^1, \mathbb{R}^d)$ Prop. 2.1.0 in [13]. Because open subsets of vector spaces are the simplest examples of manifold, $\mathrm{Imm}(S^1, \mathbb{R}^d)$ is a Fréchet manifold, modelled on the space $C^\infty(S^1, \mathbb{R}^d)$.

1.5. *The manifold* $\mathrm{Imm}_{C^n}(S^1, \mathbb{R}^d)$

Instead of smooth curves, we could consider curves belonging to some other regularity class. Let $n \geq 1$ and

$$\mathrm{Imm}_{C^n}(S^1, \mathbb{R}^d) = \{ c \in C^n(S^1, \mathbb{R}^d) : c'(\theta) \neq 0, \forall \theta \in S^1 \},$$

---

[c]This is true for $d \geq 2$; for $d = 1$ the set $\mathrm{Imm}(S^1, \mathbb{R}^d)$ is empty.

be the space of C^n-immersions. Again we need a topology on the space $C^n(S^1, \mathbb{R}^d)$. In this case

$$\|f\|_{n,\infty} = \sup_{0 \leq k \leq n} \|f^{(k)}\|_\infty\,,$$

is a norm making $(C^n(S^1, \mathbb{R}^d), \|\cdot\|_{n,\infty})$ into a Banach space.

Lemma 1.3: *For $n \geq 1$, $\mathrm{Imm}_{C^n}(S^1, \mathbb{R}^d)$ is an open subset of $C^n(S^1, \mathbb{R}^d)$.*

Proof: It is easy to see that the sets $M(f, \varepsilon, k)$ with $0 \leq k \leq n$ – after we adapt the definition of $M(f, \varepsilon, k)$ to include all functions $g \in C^n(S^1, \mathbb{R}^d)$ satisfying the inequality $\|g^{(k)} - f^{(k)}\| < \varepsilon$ – are open in $C^n(S^1, \mathbb{R}^d)$ and in the proof for $\mathrm{Imm}(S^1, \mathbb{R}^d)$ we only used $k = 1$; hence the same proof shows that $\mathrm{Imm}_{C^n}(S^1, \mathbb{R}^d)$ is open in $C^n(S^1, \mathbb{R}^d)$.   □

Thus $\mathrm{Imm}_{C^n}(S^1, \mathbb{R}^d)$ is a Banach manifold modelled on the space $C^n(S^1, \mathbb{R}^d)$. There is a connection between the spaces of C^n-immersions and the space of smooth immersions. As sets we have

$$\mathrm{Imm}(S^1, \mathbb{R}^d) = \bigcap_{n \geq 1} \mathrm{Imm}_{C^n}(S^1, \mathbb{R}^d)\,.$$

However, more is true: we can consider the diagram

$$C^\infty(S^1, \mathbb{R}^d) \subseteq \cdots \subseteq C^n(S^1, \mathbb{R}^d) \subseteq C^{n-1}(S^1, \mathbb{R}^d) \subseteq \cdots \subseteq C^1(S^1, \mathbb{R}^d)\,,$$

and topologically $C^\infty(S^1, \mathbb{R}^d) = \lim_{n \to \infty} C^n(S^1, \mathbb{R}^d)$ is the projective limit of the spaces $C^n(S^1, \mathbb{R}^d)$ [10].

1.6. *Calculus in Banach spaces*

Having chosen a modelling space E, we look back at the definition of a manifold and see that 1.1(3) requires chart changes to be smooth maps. To a considerable extent multivariable calculus generalizes without problems from a finite-dimensional Euclidean space to Banach spaces, but not beyond.

For example, if X, Y are Banach spaces and $f : X \to Y$ a function, we can define the derivative $Df(x)$ of f at $x \in X$ to be the linear map $A \in L(X, Y)$, such that

$$\lim_{h \to 0} \frac{\|f(x + h) - f(x) - A.h\|_Y}{\|h\|_X} = 0\,.$$

I want to emphasize in particular two theorems, that are valid in Banach spaces, but fail for Fréchet spaces: the existence theorem for ODEs and the inverse function theorem. First, the local existence theorem for ODEs with Lipschitz right hand sides.

Theorem 1.4: *Let X be a Banach space, $U \subseteq X$ an open subset and $F : (a,b) \times U \to X$ a continuous function, that is Lipschitz continuous in the second variable, i.e.,*

$$\|F(t,x) - F(t,y)\|_X \le C\|x - y\|_X ,$$

for some $C > 0$ and all $t \in (a,b)$, $x,y \in U$. Then, given $(t_0, x_0) \in (a,b) \times U$, there exists $x : (t_0 - \varepsilon, t_0 + \varepsilon) \to X$, such that

$$\partial_t x(t) = F(t, x(t)), \quad x(t_0) = x_0 .$$

In fact more can be said: the solution is as regular as the right hand side; if the right hand side depends smoothly on some parameters, then so does the solution; furthermore, one can estimate the length of the interval of existence. One can also get by with less regularity of $F(t,x)$ in the t-variable. This is used in the LDDMM framework [28], where vector fields $F(t,x)$ are assumed to be only integrable in the t-variable but not necessarily continuous.

For $x \in X$, denote by $B_r(x) = \{y \in X : \|y - x\|_X < r\}$ the open r-ball. The following is a version of the inverse function theorem.

Theorem 1.5: *Let X, Y be Banach spaces, $U \subseteq X$ open, $f \in C^1(U, Y)$ and $Df(x_0)$ invertible for $x_0 \in U$. Then there exists $r > 0$ such that $f(B_r(x_0))$ is open in Y and $f : B_r(x_0) \to f(B_r(x_0))$ is a diffeomorphism.*

Both of these theorems are not valid in Fréchet spaces with counterexamples given in 1.7. But even in Banach spaces life is not as easy as it was in finite dimensions. Two things are lost: local compactness and uniqueness of the topology. In fact the only locally compact vector spaces are finite-dimensional ones. The following is Thm. 1.22 in [25].

Theorem 1.6: *Let X be a topological vector space. If X has an open set, whose closure is compact, then X is finite-dimensional.*

In finite dimensions we do not have to choose, which topology to consider: there is only one n-dimensional vector space. The following is Thm. 1.21 in [25].

Theorem 1.7: *Let X be a topological vector space. If $\dim X = n$, then X is homeomorphic to $\mathbb{R}^n$.*

This is lost in infinite dimensions. There is some limited variant of the uniqueness for Banach spaces. The following can be found in [27].

Theorem 1.8: *Let $(X, \mathcal{F})$ be a topological vector space. Then, up to equivalence of norms, there is at most one norm $\|\cdot\|$ one can place on X, such that $(X, \|\cdot\|)$ is a Banach space whose topology is at least as strong as $\mathcal{F}$. In particular, there is at most one topology stronger than $\mathcal{F}$ that comes from a Banach space norm.*

1.7. *Counterexamples in Fréchet spaces*

For an example, that differential equations may not have solutions in Fréchet spaces, consider $C(\mathbb{R})$, the space of continuous functions, with the compact open topology (Sect. 2.1 in [13]). A basis for the compact open topology consists of sets

$$M(K, V) = \{f \in C(\mathbb{R}) \, : \, f(K) \subseteq V\} \, ,$$

where $K \subseteq \mathbb{R}$ is compact and $V \subseteq \mathbb{R}$ is open. The differential equation

$$\partial_t f = f^2 \, , \quad f(0, x) = x \, ,$$

has a smooth right hand side, but admits no solution in $C(\mathbb{R})$: if we look at the pointwise solution, we have

$$f(t, x) = \frac{x}{1 - tx} \, ,$$

provided $tx < 1$. Hence for no $t \neq 0$ do we obtain a function, defined on all of $\mathbb{R}$.

For an example, that the inverse function theorem fails for Fréchet spaces, consider the map

$$F : C(\mathbb{R}) \to C(\mathbb{R}), \quad f \mapsto e^f \, ,$$

where $C(\mathbb{R})$ carries the compact open topology. Its derivative is $DF(f).h = e^f.h$, which is invertible everywhere. The image of F consists of everywhere positive functions,

$$F(C(\mathbb{R})) = \{f \in C(\mathbb{R}) \, : \, f > 0\} \, ,$$

but this set is not open in the compact open topology, because the topology is not strong enough to control the behaviour towards infinity.

1.8. *Banach and Fréchet manifolds*

Why are Banach or even Hilbert manifolds not enough? One of the important objects in computational anatomy and shape analysis is the diffeomorphism group

$$\mathrm{Diff}(M) = \{\varphi \in C^\infty(M, M) : \varphi \text{ bijective, } \varphi^{-1} \in C^\infty(M, M)\},$$

of a compact manifold. We will see later that $\mathrm{Diff}(M)$ is a smooth Fréchet–Lie group. What about a Banach manifold version of the diffeomorphism group? If $n \geq 1$, then one can consider

$$\mathrm{Diff}_{C^n}(M) = \{\varphi \in C^n(M, M) : \varphi \text{ bijective, } \varphi^{-1} \in C^n(M, M)\},$$

the group of C^n-diffeomorphisms. The space $\mathrm{Diff}_{C^n}(M)$ is a Banach manifold and a topological group, but not a Lie group. What went wrong? The group operations are continuous, but not differentiable. Fix $\varphi \in \mathrm{Diff}_{C^n}(M)$ and consider the map

$$L_\varphi : \mathrm{Diff}_{C^n}(M) \to \mathrm{Diff}_{C^n}(M), \quad \psi \mapsto \varphi \circ \psi;$$

its derivative should be

$$T_\psi L_\varphi.h = (D\varphi \circ \psi).h,$$

with $T_\psi L_\varphi$ denoting the derivative of the map L_ψ and $D\varphi$ denotes the derivative of the diffeomorphism φ; the former is a map between infinite-dimensional manifolds, while the latter maps M to itself. To see this, consider a one-parameter variation $\psi(t, x)$, such that $\psi(0, x) = \psi(x)$ and $\partial_t \psi(t, x)|_{t=0} = h(x)$, and compute

$$\partial_t \varphi(\psi(t, x))|_{t=0} = D\varphi(\psi(x)).h(x).$$

We see that in general $T_\psi L_\varphi.h$ lies only in C^{n-1}. However, if composition were to be a differentiable operator, $T_\varphi L_\psi$ would have to map into C^n-functions.

There seems to be a trade-off involved: we can consider smooth functions, in which case the diffeomorphism group is a Lie group, but can be modelled only on a Fréchet space; or we look at functions with finite regularity, but then composition ceases to be differentiable. There is a theorem by Omori [22] stating that this choice cannot be avoided.

Theorem 1.9: *If a connected Banach–Lie group G acts effectively, transitively and smoothly on a compact manifold, then G must be a finite-dimensional Lie group.*

A smooth action of a Lie group G on a manifold M is a smooth map $G \times M \to M$, written as $(g, x) \mapsto g.x$, satisfying the identities $e.x = x$ and $g.(h.x) = (gh).x$, for all $g, h \in G$ and $x \in M$ with $e \in G$ the identity element (Sect. 6 in [18]). The action is called *transitive*, if for any two points $x, y \in M$ there exists $g \in G$ with $g.x = y$; the action is called *effective*, if

$$g.x = h.x \text{ for all } x \in M \Rightarrow g = h.$$

In other words, an effective action allows us to distinguish group elements based on their action on the space.

The action of the diffeomorphism group on the base manifold is given by $\varphi.x = \varphi(x)$; it is clearly effective, since $\varphi(x) = \psi(x)$ for all $x \in M$ implies $\varphi = \psi$ as functions. The diffeomorphism group also acts transitively ((43.20) in [16]).

Therefore Omori's theorem requires us to make a choice: either our diffeomorphism group is not a Banach manifold or it is not a smoothly acting Lie group, i.e., the group operations or the action on the manifold are not smooth. Choosing to work with the group $\text{Diff}(M)$ of smooth diffeomorphisms leads to the Fréchet manifold setting, where $\text{Diff}(M)$ is a Lie group; it is easier to do geometry, since more operations are differentiable, but establishing analytic results is more challenging. The other choice is a group like $\text{Diff}_{C^n}(M)$ of diffeomorphisms with finitely many derivatives. This group is a Banach manifold and hence one has multiple tools available to prove existence results; however, because $\text{Diff}_{C^n}(M)$ is not a Lie group, it is a less rich geometric setting. One can use the intuition and the language of differential geometry, but not necessarily its tools.

2. Riemannian geometry in infinite dimensions

Most examples of spaces of maps discussed in the first lecture were open subsets of vector spaces. For example, $\text{Imm}(S^1, \mathbb{R}^d)$ is an open subset of $C^\infty(S^1, \mathbb{R}^d)$ and as such its manifold structure is trivial. The second lecture begun by discussing how to define a manifold structure on nonlinear spaces of functions such as $C^\infty(M, N)$. It then proceeded to consider Riemannian metrics on infinite-dimensional manifolds with special emphasis on weak Riemannian metrics. These are Riemannian metrics on Banach and Fréchet manifolds.

2.1. *The manifold $C^\infty(M, N)$*

We assume that M is a compact (hence finite-dimensional) manifold without boundary, while N can be a noncompact (with some mild restrictions even an infinite-dimensional) manifold, also without boundary. How do we model $C^\infty(M, N)$ as an infinite-dimensional manifold? A local deformation h of a function $f \in C^\infty(M, N)$ is a vector field in N along M,

$$\begin{array}{ccc} & TN & \\ {}^{h}\nearrow & \Big\downarrow \pi_N & \qquad h(x) \in T_{f(x)}N \,. \\ M \xrightarrow{\ f\ } & N & \end{array}$$

The collection of all these deformations is

$$\Gamma(f^*TN) = \{h \in C^\infty(M, TN) : \pi_N \circ h = f\}\,,$$

the space of sections of the pullback bundle f^*TN. This is a linear space.

Choose a Riemannian metric $(N, \bar{g})$ on N. The metric gives rise to an exponential map

$$\pi_N \times \exp : TN \supseteq U \to N \times N\,, \quad h_x \mapsto (y, \exp_y(h_y))\,,$$

where $h_y \in T_yN$ and $\exp_y$ is the exponential map of $\bar{g}$ at $y \in N$. It is defined on a neighborhood of the zero section, and if U is suitably small, it is a diffeomorphism onto its image. Denote the image by $V = \pi_N \times \exp(U)$. Then any function $g \in C^\infty(M, N)$, that is close enough to f in the sense that $(f(x), g(x)) \in V$ for all $x \in M$, can be represented by

$$u_f(g) : x \mapsto (\pi_N \times \exp)^{-1}(f(x), g(x))\,, \quad u_f(g) \in \Gamma(f^*TN)\,.$$

This means that we have found around each function f an open neighborhood

$$\mathcal{V}_f = \{g \in C^\infty(M, N) : (f(x), g(x)) \in V \text{ for all } x \in M\}\,,$$

and $\mathcal{V}_f$ can be mapped bijectively onto the open subset

$$\mathcal{U}_f = \{h \in \Gamma(f^*TN) : h(x) \in U \text{ for all } x \in M\}\,,$$

of the vector space $\Gamma(f^*TN)$.

Several things have been left unsaid: one has to check that this map is indeed continuous and that its inverse is continuous as well; one also has to check that the chart change maps $h \mapsto u_f \circ u_g^{-1}(h)$ are smooth as functions of the vector fields h.

One can use the same method to construct charts on the spaces $C^n(M, N)$ of functions with finitely many derivatives. The main problem is that if $f : M \to N$ is not smooth, then the pullback bundle f^*TN is not a smooth (finite-dimensional) manifold any more. To overcome this difficulty we can use that $C^\infty(M, N)$ is dense in $C^n(M, N)$ and construct charts around all smooth f. We then have to check, that these charts indeed cover all of $C^n(M, N)^{\mathrm{d}}$.

2.2. Strong and weak Riemannian manifolds

Let M be a manifold modelled on a vector space E. A *weak Riemannian metric* G is a smooth map

$$G : TM \times_M TM \to \mathbb{R},$$

satisfying

(1) $G_x(\cdot, \cdot)$ is bilinear for all $x \in M$;
(2) $G_x(h, h) \geq 0$ for all $h \in T_xM$ with equality only for $h = 0$.

This implies that the associated map

$$\check{G} : TM \to T^*M, \quad \langle \check{G}_x(h), k \rangle = G_x(h, k),$$

is injective. In finite dimensions it would follow by counting dimensions that $\check{G}$ is bijective and hence an isomorphism. In infinite dimensions this is not longer the case.

Example 2.1: Consider the space of smooth curves $\operatorname{Imm}(S^1, \mathbb{R}^d)$ with the $\operatorname{Diff}(S^1$-invariant L^2-metric)

$$G_c(h, k) = \int_{S^1} \langle h(\theta), k(\theta) \rangle |c'| \, \mathrm{d}\theta.$$

Then $\check{G}_c(h) = h.|c'|$ and the image of $T_c \operatorname{Imm}(S^1, \mathbb{R}^d) = C^\infty(S^1, \mathbb{R}^d)$ under $\check{G}_c$ is again $C^\infty(S^1, \mathbb{R}^d)$, while the dual space $T_c^* \operatorname{Imm}(S^1, \mathbb{R}^d) = \mathcal{D}'(S^1)^d$ is the space of $\mathbb{R}^d$-valued distributions.

This means that in infinite dimensions we have to distinguish between two different notions of Riemannian metrics. A *strong Riemannian metric* is required to additionally satisfy

The topology of the inner product space $(T_xM, G_x(\cdot, \cdot))$ coincides with the topology T_xM inherits from the manifold M.

---

[d]Density of $C^\infty(M, N)$ alone is not sufficient. For example $\mathbb{Q}$ is dense in $\mathbb{R}$, but we can choose a small interval around each rational number in such a way that all the intervals still miss $\sqrt{2}$.

A strong Riemannian metric implies that T_xM and hence the modelling space of M is a Hilbert space. See [15, 17] for the theory of strong Riemannian manifolds.

Example 2.2: Let H be a Hilbert space. Then the Hilbert sphere

$$S = \{x \in H \;:\; \|x\| = 1\},$$

with the induced Riemannian metric $G_x(h, k) = \langle h, k \rangle$ for $h, k \in T_xS = \{h \in H \;:\; \langle h, x \rangle = 0\}$ is a strong Riemannian manifold.

Why do we consider weak Riemannian manifolds? There are two reasons: the only strong Riemannian manifolds are Hilbert manifolds; when we want to work with the space of smooth functions, any Riemannian metric on it will be a weak one; the other reason is that some Riemannian metrics, that are important in applications (the L^2-metric on the diffeomorphism group for example) cannot be made into strong Riemannian metrics.

2.3. *Levi-Civita covariant derivative*

After defining the Riemannian metric, one of the next objects to consider is the covariant derivative. Let (M, G) be a Riemannian manifold, modelled on E and X, Y, Z vector fields on M. Assume that $M \subseteq E$ is open or that we are in a chart for M. Then the Levi-Civita covariant derivative is given by

$$\nabla_X Y(x) = DY(x).X(x) + \Gamma(x)(X(x), Y(x)),$$

where $\Gamma : M \to L(E, E; E)$ are the *Christoffel symbols* of G,

$$2G(\Gamma(X, Y), Z) = D_{.,X}G.(Y, Z) + D_{.,Y}G.(Z, X) - D_{.,Z}G.(X, Y). \quad (2.1)$$

The important part to note is that in the definition of Γ one uses the inverse of the metric. In fact we do not need to be able to always invert it, but we need to now that the right hand side of (2.1) lies in the image $\check{G}(TM)$ of the tangent bundle under $\check{G}$. For strong Riemannian metrics this is the case, but not necessarily for weak ones.

2.4. *Example. The L^2-metric*

Consider for example the space of C^1-curves

$$\mathrm{Imm}_{C^1}(S^1, \mathbb{R}^d) = \{c \in C^1(S^1, \mathbb{R}^d) \;:\; c'(\theta) \neq 0, \; \forall \theta \in S^1\},$$

with the L^2-metric

$$G_c(h, k) = \int_{S^1} \langle h(\theta), k(\theta) \rangle |c'(\theta)| \, d\theta \,.$$

Then one can calculate that

$$D_{c,l}G.(h, k) = \int_{S^1} \langle h, k \rangle \langle l', c' \rangle \frac{1}{|c'|} \, d\theta \,,$$

and we see that the right hand side of (2.1), which is $2G(\Gamma(h, k), l)$ in this notation, involves derivatives of l, while the left hand side does not. While this is not a complete proof, it shows the idea, why the L^2-metric on the space of curves with a finite number of derivatives does not have a covariant derivative.

2.5. *The geodesic equation*

The geodesic equation plays an important role in both shape analysis and computational anatomy. Informally it describes least-energy deformations of shapes or optimal paths of transformations. From a mathematical point we can write it in a coordinate-free way as

$$\nabla_{\dot c} \dot c = 0 \,,$$

and in charts it becomes

$$\ddot c + \Gamma(c)(\dot c, \dot c) = 0 \,.$$

From this it seems clear that the geodesic equation needs the covariant derivative or equivalently the Christoffel symbols to be written down. A more concise way to say that a metric does not have a covariant derivative would be to say that the geodesic equation for the metric does not exist. Now, how can an equation fail to exist? The geodesic equation corresponds to the Euler–Lagrange equation of the energy function

$$E(c) = \frac{1}{2} \int_0^1 G_c(\dot c, \dot c) \, dt \,,$$

and a geodesic is a critical point of the energy functional, restricted to paths with fixed endpoints. In a coordinate chart we can differentiate the energy functional

$$D_{c,h}E = \int_0^1 G_c(\dot c, \dot h) + \frac{1}{2} D_{c,h}G.(\dot c, \dot c) \, dt \,.$$

The steps until now can be done with any metric. What cannot always be done is to isolate h in this expression to obtain something of the form $\int_0^1 G_c(\ldots, h) \, dt$, where the ellipsis would contain the geodesic equation.

2.6. *The geodesic distance*

Let (M, G) be a (weak) Riemannian manifold and assume M is connected. For $x, y \in M$ we can define the *geodesic distance* between them as in finite dimensions,

$$\mathrm{dist}(x, y) = \inf_{\substack{c(0)=x \\ c(1)=y}} \int_0^1 \sqrt{G_c(\dot{c}, \dot{c})}\, dt\,,$$

where the infimum is taken over all smooth paths or equivalently all piecewise C^1-paths. Then dist has the following properties:

(1) $\mathrm{dist}(x, y) \geq 0$ for $x, y \in M$;
(2) $\mathrm{dist}(x, y) = \mathrm{dist}(y, x)$;
(3) $\mathrm{dist}(x, z) \leq \mathrm{dist}(x, y) + \mathrm{dist}(y, z)$.

What is missing from the list of properties?

(4) $\mathrm{dist}(x, y) \neq 0$ for $x \neq y$.

We call this last property *point-separating*[e]. It may fail to hold for weak Riemannian metrics. This is a purely infinite-dimensional phenomenon; in fact it only happens for weak Riemannian metrics and there are explicit examples of this [19, 5].

Note that vanishing of the geodesic distance does not mean that the metric itself is degenerate. In fact, if $x \neq y$ and $c : [0, 1] \to M$ is a path with $c(0) = x$ and $c(1) = y$, then we have

$$\mathrm{Len}(c) = \int_0^1 \sqrt{G_c(\dot{c}, \dot{c})}\, dt > 0\,,$$

with a strict inequality and if $\mathrm{Len}(c) = 0$ then c must be the constant path. Thus $\mathrm{dist}(x, y) = 0$ arises, because there might exist a family of paths with positive, yet arbitrary small, length connecting the given points.

What is the *topology induced by the geodesic distance*? In finite dimensions and in fact for strong manifolds we have the following theorem, see Thm. 1.9.5 in [15].

---

[e]There is a slight difference between a point-separating and a nonvanishing geodesic distance. We say that dist is *nonvanishing*, if there exist $x, y \in M$, such that $\mathrm{dist}(x, y) \neq 0$. It follows that a point-separating distance is nonvanishing, but in general not the other way around.

Theorem 2.3: *Let* (M, G) *be a strong Riemannian manifold. Then* dist *is point-separating and the topology induced by* (M, dist) *coincides with the manifold topology.*

3. Complete Riemannian manifolds and Hopf–Rinow

The third lecture discussed completeness properties and how far the theorem of Hopf–Rinow can be generalized to infinite-dimensional manifolds. The second part studied the group of diffeomorphisms of Sobolev regularity in more detail. This group is used in computational anatomy to model anatomical deformations and it is of interest to establish completeness results for Sobolev metrics on this group.

3.1. *Completeness properties*

For a Riemannian manifold (M, G) completeness can mean several things.

(1) M is *metrically complete*, meaning that (M, dist) is a complete metric space; i.e., all Cauchy sequences with respect to dist converge.
(2) M is *geodesically complete*, meaning that every geodesic can be continued for all time.
(3) Between any two points on M (in the same connected component), there *exists a length minimizing geodesic*.

In finite dimensions the *theorem of Hopf–Rinow* states that (1) and (2) are equivalent and imply (3). The only implication one has in infinite dimensions is that on a strong Riemannian manifold metric completeness implies geodesic completeness.

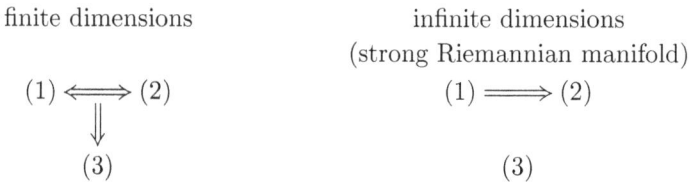

finite dimensions   infinite dimensions
(strong Riemannian manifold)

$(1) \Longleftrightarrow (2)$ $(1) \Longrightarrow (2)$
$\Downarrow$
(3) (3)

In general one cannot expect more and there are explicit counterexamples. Atkin [3] found an example of a metrically and geodesically complete manifold with two points that cannot be joined by *any* geodesic (not just a minimizing one), showing that

$$(1) \;\&\; (2) \nRightarrow (3) \,.$$

Two decades later Atkin [4] showed also that one can find a geodesically complete manifold satisfying (3), which is not metrically complete, thus showing

$$(2) \ \& \ (3) \nRightarrow (1) \,.$$

A simple example showing that metric and geodesic completeness do not imply existence of minimizing geodesics is Grossman's ellipsoid.

3.2. *Grossman's ellipsoid*

The presentation follows Sect. VIII.6 in [17]. Consider a separable Hilbert space E with an orthonormal basis $(e_n)_{n\in\mathbb{N}}$ and define the sequence $(a_n)_{n\in\mathbb{N}}$ by $a_0 = 1$ and $a_n = 1 + 2^{-n}$ for $n \geq 1$. Consider the ellipsoid

$$M = \left\{ \sum_{n\in\mathbb{N}} x_n e_n \in E \ : \ \sum_{n\in\mathbb{N}} \frac{x_n^2}{a_n^2} = 1 \right\} .$$

We can view $M = F(S)$ as the image of the unit sphere $S = \{x \in X \ : \ \|x\| = 1\}$ under the transformation

$$F : X \to X, \quad \sum_{n\in\mathbb{N}} x_n e_n \mapsto \sum_{n\in\mathbb{N}} a_n x_n e_n \,.$$

Consider a path c in S joining the two points e_0 and $-e_0$. Then Fc is a path in M joining e_0 and $-e_0$ and every path in M can be written in such a way, because F is invertible. We claim that $\mathrm{dist}(e_0, -e_0) = \pi$, but that there exists no path realizing this distance. The lengths of a path $c(t) = \sum_{n\in\mathbb{N}} c_n(t)e_n$ in S and of Fc in M are

$$\mathrm{Len}(c) = \int_0^1 \sqrt{\sum_{n\in\mathbb{N}} \dot{c}_n(t)^2} \, dt \quad \mathrm{Len}(Fc) = \int_0^1 \sqrt{\sum_{n\in\mathbb{N}} a_n^2 \dot{c}_n(t)^2} \, dt \,. \quad (3.1)$$

We certainly have

$$\pi \leq \mathrm{Len}(c) \leq \mathrm{Len}(Fc) \,,$$

since $\|\dot{c}\| \leq \|F\dot{c}\|$. In fact by looking at the sequence $(a_n)_{n\in\mathbb{N}}$ we see that $\mathrm{Len}(c) = \mathrm{Len}(Fc)$ for a curve $c(t) = \sum_{n\in\mathbb{N}} c_n(t)e_n$, if and only if $\dot{c}_n(t) = 0$ for $n \geq 1$. But the only curve in S starting at e_0 that satisfies this is the constant curve. Thus we have for c in S joining e_0 and $-e_0$ the strict inequality

$$\pi \leq \mathrm{Len}(c) < \mathrm{Len}(Fc) \,,$$

showing that $\text{dist}_M(e_0, -e_0) \geq \pi$ and $\text{Len}(Fc) > \pi$ for any curve Fc in M joining them. However, if we let c be the half great circle joining the two points in the (e_0, e_n)-plane, then

$$\text{Len}(Fc) \leq \left(1 + 2^{-n}\right)\pi \to \pi = \text{dist}_M(e_0, -e_0).$$

3.3. Sobolev spaces on $\mathbb{R}^n$

The Sobolev spaces $H^q(\mathbb{R}^d)$ with $q \in \mathbb{R}_{\geq 0}$ can be defined in terms of the Fourier transform

$$\mathcal{F}f(\xi) = (2\pi)^{-n/2} \int_{\mathbb{R}^n} e^{-i\langle x,\xi\rangle} f(x)\,\mathrm{d}x\,,$$

and consist of L^2-integrable functions f with the property that $(1 + |\xi|^2)^{q/2}\mathcal{F}f$ is L^2-integrable as well. The same definition can also be used when $q < 0$, but then we have to consider distributions f, such that $(1 + |\xi|^2)^{q/2}\mathcal{F}f$ is an L^2-integrable function. An inner product on $H^q(\mathbb{R}^d)$ is given by

$$\langle f, g\rangle_{H^q} = \mathfrak{Re} \int_{\mathbb{R}^d} (1 + |\xi|^2)^q \mathcal{F}f(\xi)\overline{\mathcal{F}g(\xi)}\,\mathrm{d}\xi\,. \tag{3.2}$$

If $q \in \mathbb{N}$, the Sobolev space $H^q(\mathbb{R}^d)$ consists of L^2-integrable functions $f : \mathbb{R}^d \to \mathbb{R}$ with the property that all distributional derivatives $\partial^\alpha f$ up to order $|\alpha| \leq q$ are L^2-integrable as well. An inner product, that is equivalent but not equal to the above is

$$\langle f, g\rangle_{H^q} = \int_{\mathbb{R}^d} f(x)g(x) + \sum_{|\alpha|=q} \partial^\alpha f(x)\partial^\alpha g(x)\,\mathrm{d}x\,. \tag{3.3}$$

Sobolev spaces satisfy the following embedding property.

Lemma 3.1: *If $q > d/2 + k$, then $H^q(\mathbb{R}^d) \hookrightarrow C_0^k(\mathbb{R}^d)$.*

In the above $X \hookrightarrow Y$ means that X is continuously embedded into Y and $C_0^k(\mathbb{R}^d)$ denotes the Banach space k-times continuously differentiable functions, that together with their derivatives vanish at infinity. When $q > d/2$, Sobolev spaces also form an algebra.

Lemma 3.2: *If $q > d/2$ and $0 \leq r \leq q$. Then pointwise multiplication can be extended to a bounded bilinear map*

$$H^q(\mathbb{R}^d) \times H^r(\mathbb{R}^d) \to H^r(\mathbb{R}^d)\,, \quad (f, g) \mapsto f \cdot g\,.$$

There are several equivalent ways to define Sobolev spaces and these definitions all lead to the same set of functions with the same topology. However, the inner products, while equivalent, are not the same. For example for $q \in \mathbb{N}$, (3.2) and (3.3) define two equivalent, but different inner products on $H^q(\mathbb{R}^d)$.

For the theory of Sobolev spaces one can consult one of the many books on the subject, e.g. [1].

3.4. *The diffeomorphism group* $\mathrm{Diff}_{H^q}(\mathbb{R}^d)$

Denote by $\mathrm{Diff}_{C^1}(\mathbb{R}^d)$ the space of C^1-diffeomorphisms of $\mathbb{R}^d$, i.e.,

$$\mathrm{Diff}_{C^1}(\mathbb{R}^d) = \{\varphi \in C^1(\mathbb{R}^d, \mathbb{R}^d) : \varphi \text{ bijective}, \varphi^{-1} \in C^1(\mathbb{R}^d, \mathbb{R}^d)\}.$$

For $q > d/2 + 1$ and $q \in \mathbb{R}$ there are three equivalent ways to define the group $\mathrm{Diff}_{H^q}(\mathbb{R}^d)$ of Sobolev diffeomorphisms:

$$\mathrm{Diff}_{H^q}(\mathbb{R}^d) = \{\varphi \in \mathrm{Id} + H^q(\mathbb{R}^d, \mathbb{R}^d) : \varphi \text{ bijective}, \varphi^{-1} \in \mathrm{Id} + H^q(\mathbb{R}^d, \mathbb{R}^d)\} \tag{3.4}$$

$$= \{\varphi \in \mathrm{Id} + H^q(\mathbb{R}^d, \mathbb{R}^d) : \varphi \in \mathrm{Diff}_{C^1}(\mathbb{R}^d)\} \tag{3.5}$$

$$= \{\varphi \in \mathrm{Id} + H^q(\mathbb{R}^d, \mathbb{R}^d) : \det D\varphi(x) > 0, \forall x \in \mathbb{R}^d\}. \tag{3.6}$$

If we denote the three sets on the right by A_1, A_2 and A_3, then it is not difficult to see the inclusions $A_1 \subseteq A_2 \subseteq A_3$. The equivalence $A_1 = A_2$ has first been shown in Sect. 3 in [9] for the diffeomorphism group of a compact manifold; a proof for $\mathrm{Diff}_{H^q}(\mathbb{R}^d)$ can be found in [14]. Regarding the inclusion $A_3 \subseteq A_2$, it is shown in [23], Corr. 4.3, that if $\varphi \in C^1$ with $\det D\varphi(x) > 0$ and $\lim_{|x|\to\infty} |\varphi(x)| = \infty$, then φ is a C^1-diffeomorphism.

It follows from the Sobolev embedding theorem, that $\mathrm{Diff}_{H^q}(\mathbb{R}^d) - \mathrm{Id}$ is an open subset of $H^q(\mathbb{R}^d, \mathbb{R}^d)$ and thus a Hilbert manifold. Since each $\varphi \in \mathrm{Diff}_{H^q}(\mathbb{R}^d)$ has to decay to the identity for $|x| \to \infty$, it follows that φ is orientation preserving, see Thm. 1.1 in [14].

Proposition 3.3: *Let $q > d/2 + 1$. Then $\mathrm{Diff}_{H^q}(\mathbb{R}^d)$ is a smooth Hilbert manifold and a topological group.*

We have the following result concerning the regularity of the composition map.

Proposition 3.4: *Let $q > d/2 + 1$ and $k \in \mathbb{N}$. Then composition*

$$\mathrm{Diff}_{H^{q+k}}(\mathbb{R}^d) \times \mathrm{Diff}_{H^q}(\mathbb{R}^d) \to \mathrm{Diff}_{H^q}(\mathbb{R}^d), \quad (\varphi, \psi) \mapsto \varphi \circ \psi,$$

and the inverse map

$$\mathrm{Diff}_{H^{q+k}}(\mathbb{R}^d) \to \mathrm{Diff}_{H^q}(\mathbb{R}^d), \quad \varphi \mapsto \varphi^{-1},$$

are C^k-maps.

This proposition means that we have to trade regularity of diffeomorphisms to obtain regularity of the composition map, i.e., if φ is of class H^{q+k}, then the composition into H^q-diffeomorphisms will be of class C^k.

We can also look at the group

$$\mathrm{Diff}_{H^\infty}(\mathbb{R}^d) = \bigcap_{q>d/2+1} \mathrm{Diff}_{H^q}(\mathbb{R}^d);$$

it consists of smooth diffeomorphisms that, together with all derivatives, decay towards infinity like L^2-functions. It is a smooth, regular, Fréchet–Lie group [20]; its Lie algebra is $\mathfrak{X}_{H^\infty}(\mathbb{R}^d) = \bigcap_{q>d/2+1} H^q(\mathbb{R}^d)$.

3.5. *Connection to LDDMM*

One beautiful property of the Sobolev diffeomorphism group is that it coincides with the group

$$\mathcal{G}_{H^q(\mathbb{R}^d,\mathbb{R}^d)} = \left\{\varphi(1) : \varphi(t) \text{ is the flow of some } u \in L^1([0,1], H^q(\mathbb{R}^d,\mathbb{R}^d))\right\},$$

which is used in the LDDMM framework. We have from Thm. 8.3 in [7].

Proposition 3.5: *Let $q > d/2 + 1$. Then*

$$\mathcal{G}_{H^q(\mathbb{R}^d,\mathbb{R}^d)} = \mathrm{Diff}_{H^q}(\mathbb{R}^d)_0,$$

where the space on the right is the connected component of Id.

Proof: Let U be a convex neighborhood around Id in $\mathrm{Diff}_{H^q}(\mathbb{R}^d)$. Then every $\psi \in U$ can be reached from Id via the smooth path $\varphi(t) = (1 - t)\,\mathrm{Id} + t\psi$. Since $\varphi(t)$ is the flow of the associated vector field $u(t) = \partial_t\varphi(t) \circ \varphi(t)^{-1}$ and $u \in C([0,1], H^q)$, it follows that $\psi \in \mathcal{G}_{H^q}$. Thus $U \subseteq \mathcal{G}_{H^q}$ and since $\mathcal{G}_{H^q}$ is a group, the same holds also for the whole connected component containing U. This shows the inclusion $\mathrm{Diff}_{H^q}(\mathbb{R}^d)_0 \subseteq \mathcal{G}_{H^q}$.

For the inclusion $\mathcal{G}_{H^q} \subseteq \mathrm{Diff}_{H^q}(\mathbb{R}^d)$ we have to show that given a vector field $u \in L^1([0,1], H^q(\mathbb{R}^d,\mathbb{R}^d))$ the flow defined by $\partial_t\varphi(t) = u(t) \circ \varphi(t)$ is a curve in $\mathrm{Diff}_{H^q}(\mathbb{R}^d)$. This is the content of [7]. $\square$

4. Riemannian metrics induced by the diffeomorphism group

The last lecture considered the action of the diffeomorphism group $\mathrm{Diff}(\mathbb{R}^d)$ on the space of embeddings $\mathrm{Emb}(M, \mathbb{R}^d)$ given by composition, $(\varphi, q) \mapsto \varphi \circ q$. Given a Riemannian metric on $\mathrm{Diff}(\mathbb{R}^d)$ as is the case in the LDDMM framework in computational anatomy this action can be used to induce a Riemannian metric on $\mathrm{Emb}(M, \mathbb{R}^d)$ such that for a given $q_0 \in \mathrm{Emb}(M, \mathbb{R}^d)$ the projection $\varphi_{q_0}(\varphi) = \varphi \circ q_0$ is a Riemannian submersion. This lecture looks at this construction and properties of the induced Riemannian metric. Some more recent analytical results can be found in [6].

4.1. The space $\mathrm{Emb}(M, \mathbb{R}^d)$

Let M be a compact manifold without boundary. We denote the space of embeddings of M into $\mathbb{R}^d$ by

$$\mathrm{Emb}(M, \mathbb{R}^d) = \{q \in C^\infty(M, \mathbb{R}^d) : q \text{ is an embedding}\};$$

to be more precise an embedding q is an immersion ($T_x q$ is injective for all $x \in M$) and a homeomorphism onto its image. It is an open subset of the space of immersions, $\mathrm{Imm}(M, \mathbb{R}^d)$, and thus also of $C^\infty(M, \mathbb{R}^d)$; hence it is a Fréchet manifold.

The shape space of embeddings is

$$B_e(M, \mathbb{R}^d) := \mathrm{Emb}(M, \mathbb{R}^d) / \mathrm{Diff}(M).$$

It can be identified with the set of all embedded submanifolds of $\mathbb{R}^d$, that are diffeomorphic to M. Regarding its manifold structure we have the following theorem from Thm. 1.5 in [8].

Theorem 4.1: *The quotient space $B_e(M, \mathbb{R}^d)$ is a smooth Hausdorff manifold and the projection*

$$\pi : \mathrm{Emb}(M, \mathbb{R}^d) \to B_e(M, \mathbb{R}^d)$$

is a smooth principal fibration with $\mathrm{Diff}(M)$ as structure group.

When $\dim M = d - 1$ and M is orientable we can define a chart around $\pi(q) \in B_e(M, \mathbb{R}^d)$ with $q \in \mathrm{Emb}(M, \mathbb{R}^d)$ by

$$\pi \circ \psi_q : C^\infty(M, (-\varepsilon, \varepsilon)) \to B_e(M, \mathbb{R}^d),$$

with ε sufficiently small, where $\psi_q : C^\infty(M, (-\varepsilon, \varepsilon)) \to \mathrm{Emb}(M, \mathbb{R}^d)$ is defined by $\psi_q(a) = q + a n_q$ and n_q is a unit-length normal vector field to q.

4.2. *Quotient representations of* $B_e(M, \mathbb{R}^d)$

Consider[f] $B_e(M, \mathbb{R}^d)$ as the space of embedded type M submanifolds of $\mathbb{R}^d$. We assume dim $M < d$ for the space to be nonempty. For the remainder of this lecture we will write $\mathrm{Diff}(\mathbb{R}^d)$ for the group[g] $\mathrm{Diff}_{H^\infty}(\mathbb{R}^d)$, in fact we only use the connected component of the identity of $\mathrm{Diff}_{H^\infty}(\mathbb{R}^d)$, and $\mathfrak{X}(M)$ for the corresponding space $\mathfrak{X}_{H^\infty}(\mathbb{R}^d)$ of vector fields.

The natural action of $\mathrm{Diff}(\mathbb{R}^d)$ on $B_e(M, \mathbb{R}^d)$ is given by

$$\mathrm{Diff}(\mathbb{R}^d) \times B_e(M, \mathbb{R}^d) \ni (\varphi, Q) \mapsto \varphi(Q) \in B_e(M, \mathbb{R}^d) \,.$$

This action is in general not transitive – consider for example a knotted and an unknotted loop in $\mathbb{R}^3$ – but it is locally transitive and hence its orbits are open subsets of $B_e(M, \mathbb{R}^d)$. Since the group $\mathrm{Diff}(\mathbb{R}^d)$ is connected, orbits of the $\mathrm{Diff}(\mathbb{R}^d)$-action are the connected components of $B_e(M, \mathbb{R}^d)$. For $Q \in B_e(M, \mathbb{R}^d)$ the isotropy group

$$\mathrm{Diff}(\mathbb{R}^d)_Q = \{\varphi \, : \, \varphi(Q) = Q\} \,,$$

consists of all diffeomorphisms that map Q to itself. Thus each orbit $\mathrm{Orb}(Q) = \mathrm{Diff}(\mathbb{R}^d).Q$ can be identified with the quotient

$$B_e(M, \mathbb{R}^d) \supseteq \mathrm{Orb}(Q) \cong \mathrm{Diff}(\mathbb{R}^d)/\mathrm{Diff}(\mathbb{R}^d)_Q \,.$$

Let us take a step backwards and remember that we defined $B_e(M, \mathbb{R}^d)$ to be the quotient

$$B_e(M, \mathbb{R}^d) \cong \mathrm{Emb}(M, \mathbb{R}^d)/\mathrm{Diff}(M) \,.$$

The diffeomorphism group $\mathrm{Diff}(\mathbb{R}^d)$ also acts on the space $\mathrm{Emb}(M, \mathbb{R}^d)$ of embeddings – i.e., the space of parametrized submanifolds – with the action

$$\mathrm{Diff}(\mathbb{R}^d) \times \mathrm{Emb}(M, \mathbb{R}^d) \ni (\varphi, q) \mapsto \varphi \circ q \in \mathrm{Emb}(M, \mathbb{R}^d).$$

This action is generally not transitive either, but has open orbits as before. For fixed $q \in \mathrm{Emb}(M, \mathbb{R}^d)$, the isotropy group

$$\mathrm{Diff}(\mathbb{R}^d)_q = \{\varphi \, : \, \varphi|q(M) \equiv \mathrm{Id}\} \,,$$

consists of all diffeomorphisms that fix the image $q(M)$ pointwise. Note the subtle difference between the two groups $\mathrm{Diff}(\mathbb{R}^d)_q$ and $\mathrm{Diff}(\mathbb{R}^d)_Q$, when $Q = q(M)$. The former consists of diffeomorphisms that fix $q(M)$ pointwise,

---

[f]The presentation follows [5], Sect. 8.

[g]Since we are acting on embeddings of a compact manifold one could equally use $\mathrm{Diff}_c(\mathbb{R}^d)$, the group of compactly supported diffeomorphisms, or $\mathrm{Diff}_S(\mathbb{R}^d)$, diffeomorphisms that decay rapidly towards the identity.

while elements of the latter only fix $q(M)$ as a set. As before we can identify each orbit $\mathrm{Orb}(q) = \mathrm{Diff}(\mathbb{R}^d).q$ with the set

$$\mathrm{Emb}(M, \mathbb{R}^d) \supseteq \mathrm{Orb}(q) \cong \mathrm{Diff}(\mathbb{R}^d)/\mathrm{Diff}(\mathbb{R}^d)_q\,.$$

The isotropy groups are subgroups of each other

$$\mathrm{Diff}(\mathbb{R}^d)_q \trianglelefteq \mathrm{Diff}(\mathbb{R}^d)_Q \leq \mathrm{Diff}(\mathbb{R}^d)\,,$$

with $\mathrm{Diff}(\mathbb{R}^d)_q$ being a normal subgroup of $\mathrm{Diff}(\mathbb{R}^d)_Q$. Their quotient can be identified with

$$\mathrm{Diff}(\mathbb{R}^d)_Q/\mathrm{Diff}(\mathbb{R}^d)_q \cong \mathrm{Diff}(M)\,.$$

Now we have the two-step process,

$$\mathrm{Diff}(\mathbb{R}^d) \to \mathrm{Diff}(\mathbb{R}^d)/\mathrm{Diff}(\mathbb{R}^d)_q \cong \mathrm{Orb}(q) \subseteq \mathrm{Emb}(M, \mathbb{R}^d)$$
$$\to \mathrm{Emb}(M, \mathbb{R}^d)/\mathrm{Diff}(M) \cong B_e(M, \mathbb{R}^d)\,.$$

In particular the open subset $\mathrm{Orb}(Q)$ of $B_e(M, \mathbb{R}^d)$ can be represented as any of the quotients

$$\mathrm{Orb}(Q) \cong \mathrm{Orb}(q)/\mathrm{Diff}(M)$$
$$\cong \mathrm{Diff}(\mathbb{R}^d)/\mathrm{Diff}(\mathbb{R}^d)_q \bigg/ \mathrm{Diff}(\mathbb{R}^d)_Q/\mathrm{Diff}(\mathbb{R}^d)_q \cong \mathrm{Diff}(\mathbb{R}^d)/\mathrm{Diff}(\mathbb{R}^d)_Q\,.$$

4.3. *Metrics induced by* $\mathrm{Diff}_{H^\infty}(\mathbb{R}^d)$

Let a right-invariant Riemannian metric G^{Diff} on $\mathrm{Diff}(\mathbb{R}^d)$ be given. Our goal is to define a metric on $\mathrm{Emb}(M, \mathbb{R}^d)$ in the following way: fix an embedding $q_0 \in \mathrm{Emb}(M, \mathbb{R}^d)$ and consider some other embedding $q = \varphi \circ q_0$ in the orbit of q_0. Define the (semi-)norm of a tangent vector $h \in T_q \mathrm{Emb}(M, \mathbb{R}^d)$ by

$$G_q^{\mathrm{Emb}}(h, h) = \inf_{X_\varphi \circ q_0 = h} G_\varphi^{\mathrm{Diff}}(X_\varphi, X_\varphi)\,,$$

with $X_\varphi \in T_\varphi \mathrm{Diff}(\mathbb{R}^d)$. Intuitively we define the length of a tangent vector $h \in T_q \mathrm{Emb}(M, \mathbb{R}^d)$ as the smallest length of a tangent vector X_φ inducing this infinitesimal deformation. If π_{q_0} is the projection

$$\pi_{q_0} : \mathrm{Diff}(\mathbb{R}^d) \to \mathrm{Emb}(M, \mathbb{R}^d)\,, \quad \pi_{q_0}(\varphi) = \varphi \circ q_0\,,$$

then

$$h = X_\varphi \circ q_0 = T_\varphi \pi_{q_0}.X_\varphi\,,$$

and the equation defining G^{Emb} is the relation between two metrics that are connected by a Riemannian submersion,

$$G_q^{\mathrm{Emb}}(h,h) = \inf_{T_\varphi \pi_{q_0} \cdot X_\varphi = h} G_\varphi^{\mathrm{Diff}}(X_\varphi, X_\varphi).$$

In fact the construction of G_q^{Emb} depends neither on the diffeomorphism φ nor on the fixed embedding q_0. To see this note that

$$X_\varphi \circ q_0 = X_\varphi \circ \varphi^{-1} \circ \varphi \circ q_0 = \left(X_\varphi \circ \varphi^{-1}\right) \circ q,$$

and $X_\varphi \circ \varphi^{-1} \in T_{\mathrm{Id}} \operatorname{Diff}(\mathbb{R}^d)$. Hence we can write

$$G_q^{\mathrm{Emb}}(h,h) = \inf_{X \circ q = h} G_{\mathrm{Id}}^{\mathrm{Diff}}(X,X), \qquad (4.1)$$

with $X \in T_{\mathrm{Id}} \operatorname{Diff}(\mathbb{R}^d)$. And this last equation depends only on q and h.

One can show that the G_q^{Emb} defined in this way is a positive semidefinite bilinear form. What is not obvious is that G_q^{Emb} depends continuously or smoothly on q. This property and that it is positive definite, have to be checked in each example.

Assuming that this construction yields a (smooth) Riemannian metric on the space $\operatorname{Emb}(M, \mathbb{R}^d)$, then this metric is invariant under $\operatorname{Diff}(\mathbb{R}^d)$, because the left-action by $\operatorname{Diff}(\mathbb{R}^d)$ commutes with the right-action by $\operatorname{Diff}(M)$:

$$G_{q \circ \varphi}^{\mathrm{Emb}}(h \circ \varphi, h \circ \varphi) = \inf_{X \circ q \circ \varphi = h \circ \varphi} G_{\mathrm{Id}}^{\mathrm{Diff}}(X,X) = \inf_{X \circ q = h} G_{\mathrm{Id}}^{\mathrm{Diff}}(X,X) = G_q^{\mathrm{Emb}}(h,h).$$

The metric G^{Emb} then can be expected to project to a Riemannian metric on $B_e(M, \mathbb{R}^d)$.

4.4. *Existence of optimal lifts*

Consider the metric

$$G_{\mathrm{Id}}^{\mathrm{Diff}}(X,Y) = \int_{\mathbb{R}^d} \langle (\mathrm{Id} - \Delta)^n X, Y \rangle \, \mathrm{d}x,$$

and set $LX = (\mathrm{Id} - \Delta)^n$. For $h \in T_q \operatorname{Emb}(M, \mathbb{R}^d)$, how should an $X \in \mathfrak{X}(\mathbb{R}^d)$ satisfying $X \circ q = h$ and

$$G_q^{\mathrm{Emb}}(h,h) = G_{\mathrm{Id}}^{\mathrm{Diff}}(X,X)$$

look like? It has to satisfy $G_{\mathrm{Id}}^{\mathrm{Diff}}(X,Y) = 0$ for all Y with $Y \circ q \equiv 0$. In other words

$$\int_{\mathbb{R}^d} \langle LX, Y \rangle \, \mathrm{d}x = 0, \quad \forall Y \in \mathfrak{X}(\mathbb{R}^d) \text{ with } Y \circ q \equiv 0.$$

Because $q(M)$ is a set of positive codimension and hence zero measure, there exists *no smooth* function LX satisfying this and therefore there exists no smooth X attaining the infimum in (4.1). To find an infimum we have to look in a bigger space of less regular functions, for example we have hope to succeed if we allow LX to be a distribution supported on the set $q(M)$.

4.5. *The RKHS point of view*

Let $(\mathcal{H}, \langle \cdot, \cdot \rangle_{\mathcal{H}})$ be a Hilbert space of vector fields, such that the canonical inclusions in the following diagram

$$\mathfrak{X}_{H^\infty}(\mathbb{R}^d) \hookrightarrow \mathcal{H} \hookrightarrow C_0^k(\mathbb{R}^d, \mathbb{R}^d)$$

are bounded linear maps and $\mathfrak{X}_{H^\infty}(\mathbb{R}^d)$ is dense in $\mathcal{H}$. We say that $\mathcal{H}$ is *k-admissible*, if the inclusion $\mathcal{H} \hookrightarrow C_0^k$ is bounded. The motivation for the notion of k-admissible spaces of vector fields and their use to define groups of diffeomorphisms is explained in [28].

The induced right-invariant metric on $\mathrm{Diff}(\mathbb{R}^d)$ is

$$G_\varphi^{\mathrm{Diff}}(X_\varphi, Y_\varphi) = \langle X_\varphi \circ \varphi^{-1}, Y_\varphi \circ \varphi^{-1} \rangle_{\mathcal{H}},$$

and the metric on $\mathrm{Emb}(M, \mathbb{R}^d)$ is

$$G_q^{\mathrm{Emb}}(h, h) = \inf_{X \circ q = h} \langle X, X \rangle_{\mathcal{H}}.$$

Lemma 4.2: *If the vector space $(\mathcal{H}, \langle \cdot, \cdot \rangle_{\mathcal{H}})$ is 0-admissible, then the induced metric G^{Emb} on $\mathrm{Emb}(M, \mathbb{R}^d)$ is nondegenerate.*

Proof: Let $h \in T_q \mathrm{Emb}(M, \mathbb{R}^d)$ and $x \in M$, such that $h(x) \neq 0$. Then

$$h(x) \leq \|h\|_\infty = \|X \circ q\|_\infty \leq \|X\|_\infty \leq C \sqrt{\langle X, X, \rangle_{\mathcal{H}}}.$$

Since this holds for all $X \in \mathfrak{X}(\mathbb{R}^d)$ with $X \circ q = h$, we conclude that

$$G_q^{\mathrm{Emb}}(h, h) \geq C^{-2} |h(x)|^2 > 0,$$

and hence the metric is nondegenerate. $\qquad\qquad\square$

4.6. *The horizontal subspace*

To compute an explicit expression for G_q^{Emb} we decompose $\mathcal{H}$ into

$$\mathcal{H}_q^{\mathrm{ver}} = \{X \in \mathcal{H} : X \circ q \equiv 0\}, \qquad \mathcal{H}_q^{\mathrm{hor}} = \left(\mathcal{H}_q^{\mathrm{ver}}\right)^\perp.$$

Note that since $\mathcal{H}$ is 0-admissible, $\mathcal{H}_q^{\mathrm{ver}}$ is a closed subspace and hence $\mathcal{H}_q^{\mathrm{ver}} \oplus \mathcal{H}_q^{\mathrm{hor}} = \mathcal{H}$. Then the induced metric is given by

$$G_q^{\mathrm{Emb}}(h, h) = \langle X^{\mathrm{hor}}, X^{\mathrm{hor}} \rangle_{\mathcal{H}},$$

where $X \in \mathfrak{X}(\mathbb{R}^d)$ is any vector field satisfying $X \circ q = h$ and $X^{\mathrm{hor}} \in \mathcal{H}_q^{\mathrm{hor}}$ is its horizontal projection. The horizontal projection does not depend on the choice of the lift, i.e. if $X, Y \in \mathfrak{X}(\mathbb{R}^d)$ coincide along q, then $X - Y \in \mathcal{H}_q^{\mathrm{ver}}$ and hence $X^{\mathrm{hor}} = Y^{\mathrm{hor}}$.

We have the maps

$$\begin{array}{cc}
T_q \operatorname{Emb}(M, \mathbb{R}^d) \to \mathcal{H}_q^{\mathrm{hor}} & \mathcal{H}_q^{\mathrm{hor}} \to C^k(M, \mathbb{R}^d) \\
h \qquad\quad \mapsto X^{\mathrm{hor}}, & X \quad \mapsto \quad X \circ q
\end{array}.$$

The composition of these two maps is the canonical embedding $T_q \operatorname{Emb}(M, \mathbb{R}^d) \hookrightarrow C^k(M, \mathbb{R}^d)$. Because M is compact we do not have to distinguish between C^k and C_b^k. Furthermore the equation $G_q^{\mathrm{Emb}}(h, h) = \langle X^{\mathrm{hor}}, X^{\mathrm{hor}} \rangle_{\mathcal{H}}$ shows that the first map is an isometry between $(T_q \operatorname{Emb}(M, \mathbb{R}^d), G_q^{\mathrm{Emb}})$ and $\mathcal{H}_q^{\mathrm{hor}}$.

Lemma 4.3: *The image of $T_q \operatorname{Emb}(M, \mathbb{R}^d)$ is dense in $\mathcal{H}_q^{\mathrm{hor}}$ and the G_q^{Emb}-completion of $T_q \operatorname{Emb}(M, \mathbb{R}^d)$ can be identified with $\mathcal{H}_q^{\mathrm{hor}}$.*

Proof: It is enough to show that the image is dense. Given $X \in \mathcal{H}_q^{\mathrm{hor}}$ choose a sequence $X_n \in \mathfrak{X}_{H^\infty}(\mathbb{R}^d)$ converging to X in $\mathcal{H}$. Then X_n^{hor} is the image of $X_n \circ q \in T_q \operatorname{Emb}(M, \mathbb{R}^d)$ in $\mathcal{H}_q^{\mathrm{hor}}$ and

$$\left\| X_n^{\mathrm{hor}} - X \right\|_{\mathcal{H}} = \left\| X_n^{\mathrm{hor}} - X^{\mathrm{hor}} \right\|_{\mathcal{H}} \leq \left\| X_n - X \right\|_{\mathcal{H}} \to 0.$$

Hence the image is dense. $\qquad\qquad\square$

A consequence of this lemma is that the G_q^{hor}-completion of $T_q \operatorname{Emb}(M, \mathbb{R}^d)$ can be identified with a closed subspace of a RKHS and as such it is itself an RKHS. Since the norm G_q^{Emb} is defined using the infimum $G_q^{\mathrm{Emb}}(h, h) = \langle X^{\mathrm{hor}}, X^{\mathrm{hor}} \rangle_{\mathcal{H}}$, it follows from [2], Thm. I.5, that its reproducing kernel is given by restricting the kernel K of $\mathcal{H}x$ to M; i.e.,

$$K_q : M \times M \to \mathbb{R}^{d \times d}, \quad K_q(x, y) = K(q(x), q(y)),$$

is the reproducing kernel of the induced inner product on $T_q \operatorname{Emb}(M, \mathbb{R}^d)$.

5. Diff(M) as a Lie group

Having discussed infinite-dimensional Riemannian manifolds, let us briefly look at an infinite-dimensional Lie group and some of its properties.

When M is a manifold we can consider

$$\text{Diff}(M) = \{\varphi \in C^\infty(M, M) : \varphi \text{ bijective}, \varphi^{-1} \in C^\infty(M, M)\},$$

and $\mathfrak{X}(M)$, the space of vector fields on M. Intuitively we would like to see Diff(M) as a Lie group with $\mathfrak{X}(M)$ as its Lie algebra and the Lie group exponential map being the time 1 flow map of the vector field,

$$\exp : \mathfrak{X}(M) \to \text{Diff}(M), \quad u \mapsto \varphi(1),$$

where $\varphi(t)$ is the solution of the ODE

$$\partial_t \varphi(t, x) = u(\varphi(t, x)), \quad \varphi(0, x) = x.$$

This runs into several difficulties. First, exp *might not be well-defined.* Consider $M = \mathbb{R}$ and the vector field

$$u(x) = x^2 \frac{\partial}{\partial x}.$$

Its flow is given by

$$\varphi(t, x) = \frac{x}{1 - tx},$$

and we see that the vector field u is not complete. As a consequence $\varphi(1, x)$ is defined only for $x < 1$. This problem can be avoided either by restricting to compact manifolds M or by requiring vector fields to decay sufficiently rapidly towards infinity. But even then the exponential map exhibits some unexpected behaviour.

The exponential map *is not locally surjective.* Consider an element $\varphi \in$ Diff(S^1), such that φ has no fixed points and at least one isolated periodic point[h]. Assume that we can write $\varphi = \exp(u)$ for some $u \in \mathfrak{X}(S^1)$. Then u must satisfy $u(x) \neq 0$ for all $x \in S^1$. Surprisingly, we can now show that φ is conjugate to a rotation. To see this define the diffeomorphism

$$\eta(x) = c \int_0^x \frac{dy}{u(y)}, \quad c = 2\pi \left(\int_{S^1} \frac{dx}{u(x)} \right)^{-1}.$$

It is easy to check that $\eta \circ \varphi \circ \eta^{-1}$ is a rotation by calculating the derivative $\partial_t \left(\eta \circ \exp(tu) \circ \eta^{-1} \right)$. Now let $\eta \circ \varphi = R_\alpha \circ \eta$, where $R_\alpha(x) = x + \alpha \mod 2\pi$

---

[h]A point $x \in S^1$, such that $\varphi^n(x) = x$ for some $n \in \mathbb{N}_{>0}$.

is a rotation. Then $\eta \circ \varphi^n = R_{n\alpha} \circ \eta$. Let x_0 be the isolated periodic point such that $\varphi^n(x_0) = x_0$. Then $\eta(x_0) = R_{n\alpha}(\eta(x_0))$, which implies that $R_{n\alpha} = \text{Id}_{S^1}$ and thus $\varphi^n = \text{Id}_{S^1}$, which contradicts the assumption that x_0 is an isolated periodic point. Thus φ cannot be written as $\varphi = \exp(u)$ for any $u \in \mathfrak{X}(S^1)$. An example of such a φ is given by

$$\varphi(x) = x + \frac{2\pi}{n} + \varepsilon \sin(nx),$$

where we can choose $n \in \mathbb{N}$ and $|\varepsilon| < 2/n$. In particular, by choosing n large and ε small, φ will be arbitrary close to the identity. This counterexample can be found in [12], I.5.5.2, and [21], p.1017. One can show more, see [11].

Theorem 5.1: *Given a C^n-manifold M, there exists a continuous curve $\gamma : [0,1) \to \text{Diff}_c^n(M)$, $\gamma(0) = \text{Id}$, such that $\{\gamma(t) : t \in (0,1)\}$ is a set of free generators of a subgroup of $\text{Diff}_c^n(M)$, which contains only (apart from the identity) diffeomorphisms that are not in the image of the exponential map.*

Here $\text{Diff}_c^k(M)$ denotes the group of *compactly supported C^n-diffeomorphisms*, i.e., φ is compactly supported if the set $\{x \in M : \varphi(x) \neq x\}$ has compact closure.

Finally the exponential map is *not locally injective* either. Let $\psi \in \text{Diff}(S^1)$ be a $2\pi/n$-periodic diffeomorphism, i.e., $\psi(x + 2\pi/n) = \psi(x) + 2\pi/n$. Denote by $R_\alpha(x) = x + \alpha \mod 2\pi$ the rotation by α. Then $R_{2\pi t/n}$ lies in the 1-parameter subgroup $\varphi(t) = \psi \circ R_{2\pi t/n} \circ \psi^{-1}$. In other words, define the vector field

$$u(x) = \frac{2\pi}{n} \psi'(\psi^{-1}(x)) \frac{\partial}{\partial x}.$$

Its flow is

$$\varphi(t, x) = \psi \left(\psi^{-1}(x) + \frac{2\pi}{n} t \right),$$

and $\exp(u) = \varphi(1) = R_{2\pi t/n}$. Since ψ can be chosen to be arbitrary close to the identity, exp cannot be locally injective.

References

1. Robert A. Adams. *Sobolev Spaces*. Academic Press, 2nd edition edition, 2003.
2. N. Aronszajn. Theory of reproducing kernels. *Trans. Amer. Math. Soc.*, 68:337–404, 1950.
3. C. J. Atkin. The Hopf–Rinow theorem is false in infinite dimensions. *Bull. London Math. Soc.*, 7(3):261–266, 1975.

4. Christopher J. Atkin. Geodesic and metric completeness in infinite dimensions. *Hokkaido Math. J.*, 26(1):1–61, 1997.
5. Martin Bauer, Martins Bruveris, and Peter W. Michor. Overview of the geometries of shape spaces and diffeomorphism groups. *J. Math. Imaging Vis.*, 50:60–97, 2014.
6. Martins Bruveris. Riemannian geometry on spaces of submanifolds induced by the diffeomorphism group. 2017.
7. Martins Bruveris and François-Xavier Vialard. On completeness of groups of diffeomorphisms. *J. Eur. Math. Soc. (JEMS)*, 19(5):1507–1544, 2017.
8. Vicente Cervera, Francisca Mascaró, and Peter W. Michor. The action of the diffeomorphism group on the space of immersions. *Differential Geom. Appl.*, 1(4):391–401, 1991.
9. David G. Ebin. The manifold of Riemannian metrics. In *Global Analysis (Proc. Sympos. Pure Math., Vol. XV, Berkeley, Calif., 1968)*, pages 11–40. Amer. Math. Soc., Providence, R.I., 1970.
10. Paul Garrett. Functions on circles: Fourier series, I, 4 2013. http://www.math.umn.edu/~garrett/m/fun/notes_2012-13/04_blevi_sobolev.pdf.
11. Janusz Grabowski. Free subgroups of diffeomorphism groups. *Fund. Math.*, 131(2):103–121, 1988.
12. Richard S. Hamilton. The inverse function theorem of Nash and Moser. *Bull. Amer. Math. Soc. (N.S.)*, 7(1):65–222, 1982.
13. Morris W. Hirsch. *Differential topology*, volume 33 of *Graduate Texts in Mathematics*. Springer-Verlag, New York, 1994. Corrected reprint of the 1976 original.
14. H. Inci, T. Kappeler, and P. Topalov. On the regularity of the composition of diffeomorphisms. *Mem. Amer. Math. Soc.*, 226(1062):vi+60, 2013.
15. Wilhelm P. A. Klingenberg. *Riemannian Geometry*, volume 1 of *de Gruyter Studies in Mathematics*. Walter de Gruyter & Co., Berlin, second edition, 1995.
16. A. Kriegl and P. W. Michor. *The Convenient Setting of Global Analysis*, volume 53 of *Mathematical Surveys and Monographs*. American Mathematical Society, Providence, RI, 1997.
17. Serge Lang. *Fundamentals of Differential Geometry*, volume 191 of *Graduate Texts in Mathematics*. Springer-Verlag, New York, 1999.
18. P. W. Michor. *Topics in Differential Geometry*, volume 93 of *Graduate Studies in Mathematics*. American Mathematical Society, Providence, RI, 2008.
19. Peter W. Michor and David Mumford. Vanishing geodesic distance on spaces of submanifolds and diffeomorphisms. *Doc. Math.*, 10:217–245, 2005.
20. Peter W. Michor and David Mumford. A zoo of diffeomorphism groups on $\mathbb{R}^n$. *Ann. Global Anal. Geom.*, 44(4):529–540, 2013.
21. J. Milnor. Remarks on infinite-dimensional Lie groups. In *Relativity, groups and topology, II (Les Houches, 1983)*, pages 1007–1057. North-Holland, Amsterdam, 1984.
22. Hideki Omori. On Banach–Lie groups acting on finite dimensional manifolds. *Tôhoku Math. J.*, 30(2):223–250, 1978.

23. Richard S. Palais. Natural operations on differential forms. *Trans. Amer. Math. Soc.*, 92:125–141, 1959.

24. Albrecht Pietsch. *Nuclear locally convex spaces*. Springer-Verlag, New York-Heidelberg, 1972. Translated from the second German edition by William H. Ruckle, Ergebnisse der Mathematik und ihrer Grenzgebiete, Band 66.

25. Walter Rudin. *Functional Analysis*. International Series in Pure and Applied Mathematics. McGraw-Hill, Inc., New York, second edition, 1991.

26. A. Srivastava and E. Klassen. *Functional and Shape Data Analysis*. Springer Series in Statistics, 2016.

27. Terence Tao. Application of the closed graph theorem, 4 2016. http://terrytao.wordpress.com/2016/04/22/ a-quick-application-of-the-closed-graph-theorem/.

28. Laurent Younes. *Shapes and Diffeomorphisms*. Springer, 2010.

Diffeomorphic Registration of Discrete Geometric Distributions

Hsi-Wei Hsieh* and Nicolas Charon†

Center of Imaging Sciences,
Johns Hopkins University,
3400 N. Charles St, Clark Hall 317, USA
*hhsieh@cis.jhu.edu
†charon@cis.jhu.edu

This paper introduces a new framework and algorithms to address the problem of diffeomorphic registration on a general class of geometric objects that can be described as discrete distributions of local direction vectors. It builds on both the large deformation diffeomorphic metric mapping (LDDMM) model and the concept of oriented varifolds introduced in previous works like [16]. Unlike previous approaches in which varifold representations are only used as surrogates to define and evaluate fidelity terms, the specificity of this paper is to derive direct deformation models and corresponding matching algorithms for discrete varifolds. We show that it gives on the one hand an alternative numerical setting for curve and surface matching but that it can also handle efficiently more general shape structures, including multi-directional objects or multi-modal images represented as distributions of unit gradient vectors.

Contents

1. Introduction

Background

Statistical shape analysis is now regarded across the board as an important area of applied mathematics as it has been and still is the source of quantities of theoretical works as well as applications to domains like computational anatomy, computer vision or robotics. Broadly speaking, one of its central aim is to provide quantitative/computational tools to analyze the variability of geometric structures in order to perform different tasks such as shape comparison or classification.

There are several specific difficulties in tackling such problems in the case of datasets involving geometric shapes. A fundamental one is the issue of defining and computing metrics on shape spaces. A now quite standard approach which was pioneered by Grenander in [13] is to compare shapes through distances based on deformation groups equipped with right-invariant metrics together with a left group action defined on the set of shapes. In this framework, the induced distance is typically obtained by solving a registration problem i.e. by finding an optimal deformation mapping one object on the other one. It is thus ultimately determined by the deformation group and its metric for which many models have been proposed. In this paper, we will focus on the Large Deformation Diffeomorphic Metric Mapping (LDDMM) of [4] in which diffeomorphic transformations are generated as flows of time-dependent velocity fields.

Despite the versatility of such models, one of the other common difficulty in shape analysis is the multiple forms or modalities that shapes may take. Looking only at the applications in the field of computational anatomy, if early works have mostly considered shapes given by medical images [19, 4] or manually extracted landmarks [15], the variety of geometric structures at hand has considerably increased since then, whether shapes are images acquired through multiple modalities (MRI, CT...) [3], vector or tensor fields as in Diffusion Tensor Images [5], fields of orientation distribution functions [8] or delineated objects like point clouds, curves [12], surfaces [10], fiber bundles [9]...

The intent of this paper is to make a modest step toward one possible generalized setting that could encompass a rich class of shapes including many of the previous cases within a common representation and eventually lead to a common LDDMM matching framework. Our starting point is the set of works on curve and surface registration based on geometric distributions like measures, currents or varifolds [11, 10, 7]. In the recent article [16] for instance, an oriented curve/surface is interpreted as a directional distribution (known as oriented varifold) of its oriented tangent/normal vectors, which results in simple fidelity terms used in combination with LDDMM to formulate and solve inexact matching problems. Yet all those works so far have restricted the role of distributions' representations to intermediates for the computation of guiding terms in registration algorithms; the underlying deformation model and registration problem remains defined over point sets with meshes.

The stance we take here is to instead introduce group actions and formulate the diffeomorphic matching problem directly in spaces of geometric distributions. In this particular work, we will restrict the analysis to objects in 2D and 3D and focus on the simpler subspace of discrete distributions, i.e. that write as finite sums of Dirac varifold masses: Figure 1 gives a few examples of objects naturally represented in this form. We shall consider different models of group actions and derive the corresponding optimal control problems, optimality conditions (Section 3) and registration algorithms (Section 4). This provides, on the one hand, an alternative (and theoretically equivalent) numerical framework to [16] for curve and surface matching using currents, oriented or unoriented varifolds. But the main contribution of our proposed model is that it extends LDDMM registration to the more general class of objects representable by discrete varifolds. In Section 5, we will show several examples of synthetic data besides curves or surfaces that can be treated as such, including cases like multi-directional objects or contrast-invariant images.

Related works

A few past works share some close connections with the present paper. For instance, [5] develops an approach for registration of vector fields also within the LDDMM setting. The discrete distributions we consider here are however distinct from vector fields as they should rather be interpreted as unlabelled particles at some locations in space with orientation vectors attached (and with possibly varying number of orientation vectors at a single position) as opposed to a field of vectors defined on a fixed grid. In particular, our approach will be naturally framed in the Lagrangian setting as opposed to the Eulerian formulation of [5]. The deformation model and geodesic equations that are derived in Section 3 can be also related to the framework of [20] and [14] that introduce higher-order "jet" particles and higher-order similarity measures. These remain defined through labelled sets of control points though, which is again different and arguably less flexible than the approach we develop here.

2. Shapes and discrete varifolds

The idea of representing shapes as distributions goes back to the many works within the field of Geometric measure theory. Those concepts have later been of great interest in the construction of simple and numerically tractable metrics between curves or surfaces for registration problems: the works of [12, 10, 9, 7] are a few examples. The framework of oriented varifolds recently exploited in [16] was shown to encompass all those notions into a general representation and provide a wide range of metrics on the spaces of embedded curves or surfaces. We give a brief summary of the latter work below.

In the rest of the paper, we will call an oriented varifold or, to abbreviate, a *varifold* in $\mathbb{R}^n$ (we shall here consider the cases $n = 2$ or $n = 3$) a distribution on the product $\mathbb{R}^n \times \mathbb{S}^{n-1}$. In other words, a varifold μ is by definition a linear form over a certain space W of smooth functions on $\mathbb{R}^n \times \mathbb{S}^{n-1}$, which evaluation we shall write as $\mu(\omega)$ for any test function $\omega \in W$. In all what follows, we shall restrict our focus to 'discrete' shapes and varifolds, leaving aside the analysis of the corresponding continuous models. By discrete varifold, we mean specifically that μ writes as a finite combination of Dirac masses $\mu = \sum_{i=1}^{P} r_i \delta_{(x_i, d_i)}$ with $r_i > 0$, $(x_i, d_i) \in \mathbb{R}^n \times \mathbb{S}^{n-1}$ for all i, in which case $\mu(\omega) = \sum_{i=1}^{P} r_i \omega(x_i, d_i)$ for all ω. Such a μ can be thought as a set of unit direction vectors d_i located at positions x_i with weights (or masses) equal to the r_i's. We assume by convention that the (x_i, d_i) are distinct, but not necessarily that all the positions x_i are: in other words, in our model, there can be more than a single direction vector attached to each position. In the rest of the paper, we will denote by $\mathcal{D}$ the set of all discrete varifolds. Note that in this representation and unlike the cases of landmarks and vector fields, the particles are unlabelled i.e. the varifold μ is invariant to any permutation of the (x_i, d_i). One particular subset of interest that we shall denote $\overset{\circ}{\mathcal{D}} \subset \mathcal{D}$ is the space of discrete varifolds with distinct positions x_i (or equivalently, the discrete varifolds that carry a single direction vector per point position).

The relationship between shapes and varifolds relies on the fact that discrete shapes, namely curve or surface meshes, can be naturally approximated by varifolds of the previous form. As explained with more details in the aforementioned references, this is done by associating to any cell of the discrete mesh (i.e. a segment for curves or a triangular face for surfaces) the weighted Dirac $r_i \delta_{(x_i, d_i)}$ as illustrated in Figure 1. In that expression, x_i is the coordinates of the center of the

(a)   (b)

(c)   (d)

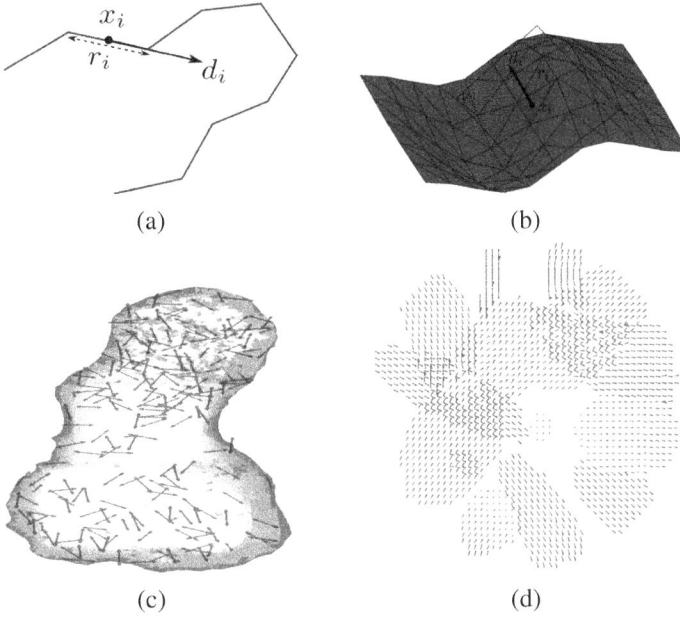

Fig. 1. Some examples of data representable by discrete varifolds: (a) Piecewise linear curve. (b) Triangulated surface. (c) A set of cells' mitosis directions measured inside a mouse embryonic heart membrane (cf. [18]). (d) Peak diffusion directions extracted from a slice of High Angular Resolution Diffusion Imaging phantom data, note the presence of multiple directions at certain locations corresponding to fiber crossing.

cell, r_i its total length or area and d_i the direction of the tangent space represented by the unit tangent or normal orientation vector d_i. It results in a mapping $S \mapsto \mu_S$ that associates to any discrete shape S the discrete varifold $\mu_S = \sum_{i=1}^{F} r_i \delta_{(x_i,d_i)} \in \overset{\circ}{\mathcal{D}}$ obtained as the sum over all faces $i = 1, \ldots, F$ of the corresponding Diracs.

The main interest of such a representation is that it gives a convenient setting for the definition of shape similarities that are easy to compute and **do not assume pairwise correspondences between points**. With the two conditions, which are quite natural in our context, that W is a Hilbert space and that all Diracs $\delta_{(x,d)}$ for $(x,d) \in \mathbb{R}^n \times \mathbb{S}^{n-1}$ belong to the dual, W then must be a Reproducing Kernel Hilbert Space (**RKHS**) associated to a smooth positive definite kernel on $\mathbb{R}^n \times \mathbb{S}^{n-1}$. In particular, we will follow the construction proposed in [16] and consider separable kernels of the form $k(x,d,x',d') = \rho(|x - x'|^2)\gamma(\langle d,d' \rangle)$ where ρ and γ define positive definite kernel functions respectively on the positions between particles and the angles between their orientation vectors. The reproducing kernel metric on W then gives a dual metric on varifolds that explicitly writes, for $\mu =$

$\sum_{i=1}^{P} r_i \delta_{(x_i,d_i)}$:

$$\|\mu\|_{W^*}^2 = \sum_{i,j} r_i r_j \rho(|x_i - x_j|^2) \gamma(\langle d_i, d_j \rangle). \tag{2.1}$$

Such metrics on W^* are determined by the choice of the positive definite functions ρ and γ and provide a global measure of proximity between two discrete varifolds. One important advantage for applications to e.g. registration is that the computation of a distance $\|\mu - \mu'\|_{W^*}^2$ between two distributions does not require finding correspondences between their masses but instead reduces numerically to a quadratic number of kernel evaluations. The gradients of the metric with respect to the x_i's and d_i's is also very easy to obtain by direct differentiation of (2.1). Finally, we note that the expression in (2.1) is also invariant to the action of the group of rigid motion. Namely for any rotation matrix R, translation vector h and the group action $(R,h) \cdot \mu \doteq \sum_{i=1}^{P} r_i \delta_{(Rx_i+h,Rd_i)}$, one has $\|(R,h) \cdot \mu\|_{W^*} = \|\mu\|_{W^*}$.

In all generality however, (2.1) may only yield a pseudo-metric on the set of discrete varifolds $\mathcal{D}$ since the inclusion mapping $\mathcal{D} \to W^*$ is not necessarily injective. A necessary and sufficient condition is:

Proposition 2.1: *The metric $\|\cdot\|_{W^*}$ on W^* induces a metric on $\mathcal{D}$ if and only if k is a strictly positive definite kernel on $\mathbb{R}^n \times \mathbb{S}^{n-1}$.*

The proof follows immediately from the definition of strictly positive definite kernel. This condition holds in particular if both kernels defined by ρ and γ are strictly positive definite. In the case of $\overset{\circ}{\mathcal{D}}$, one can provide different sufficient conditions which are often more convenient to satisfy in practice. These involve a density property on kernels called C_0-universality, cf [6]. A kernel on $\mathbb{R}^n$ is said to be C_0-universal if the associated RKHS is dense in $C_0(\mathbb{R}^n, \mathbb{R})$. Then one has the following

Proposition 2.2: *If the kernel defined by ρ is C_0-universal, $\gamma(1) > 0$ and $\gamma(u) < \gamma(1)$ for all $u \in [-1,1)$, then $\|\cdot\|_{W^*}$ induces a metric on $\overset{\circ}{\mathcal{D}}$.*

Proof: Let W_{pos} and W_{or} be the RKHS associated to ρ and γ. By contradiction, suppose that $\mu, \mu' \in \overset{\circ}{\mathcal{D}}$ with $\|\mu - \mu'\|_{W^*} = 0$ and $\mu \neq \mu'$ in $\overset{\circ}{\mathcal{D}}$. We can write μ, μ' in the following form:

$$\mu = \sum_{i=1}^{N} r_i \delta_{(z_i,d_i)}, \quad \mu' = \sum_{i=1}^{N} r_i' \delta_{(z_i,d_i')},$$

where $\{z_i\}$, with z_i all distinct, is the reunion of point positions from both distributions and $\max\limits_{1 \leq i \leq N} \{r_i, r_i'\} > 0$, $\min\limits_{1 \leq i \leq N} \{r_i, r_i'\} \geq 0$. Since μ and μ' are distinct in $\overset{\circ}{\mathcal{D}}$, there is some i_0 such that $(d_{i_0}, r_{i_0}) \neq (d'_{i_0}, r'_{i_0})$. Without loss of generality, we may assume $r_{i_0} \geq r'_{i_0}$. Let $g(\cdot) = \gamma(\langle d_{i_0}, \cdot \rangle) \in W_{or}$ and choose $f \in C_0(\mathbb{R}^n, \mathbb{R})$ satisfying $f(z_{i_0}) = 1$ and $f(z_i) = 0$ for all $i \geq 2$. Since the kernel defined by ρ is C_0-universal, there exists $\{f_n\} \subset W_{pos}$ such that $f_n \to f$ uniformly. As $f_n \otimes g \in W$, we have that

$$0 = (\mu - \mu' | f_n \otimes g) = \sum_{i=1}^{N} f_n(z_i)(r_i g(d_i) - r'_i g(d'_i)).$$

Taking the limit $n \to +\infty$, this gives:

$$0 = f(z_{i_0})(r_{i_0} g(d_{i_0}) - r'_{i_0} g(d'_{i_0})) = \underbrace{r_{i_0} \gamma(1) - r'_{i_0} \gamma(\langle d_{i_0}, d'_{i_0} \rangle)}_{A}. \qquad (2.2)$$

Since $(d_{i_0}, r_{i_0}) \neq (d'_{i_0}, r'_{i_0})$, $r_{i_0} \geq r'_{i_0}$ and $r_{i_0} > 0$, we have either $d_{i_0} \neq d'_{i_0}$ and then $A \geq r_{i_0}(\gamma(1) - \gamma(\langle d_{i_0}, d'_{i_0} \rangle)) > 0$ or $d_{i_0} = d'_{i_0}$ and $r_{i_0} > r'_{i_0}$ in which case $A = (r_{i_0} - r'_{i_0})\gamma(1) > 0$. In either case the right hand side of (2.2) is positive which is a contradiction. $\qquad \square$

Note that the C_0-universality assumption still implies that the kernel defined by ρ is strictly positive definite. However, the assumptions on γ are typically less restrictive than in Proposition 2.1.

A last subclass of varifold metrics that shall be of interest in this paper is the case of orientation-invariant kernels which amounts in choosing an even function γ in the kernel definition. This, indeed, leads to a space W^* and metric $\| \cdot \|_{W^*}$ for which Diracs $\delta_{(x,d)}$ and $\delta_{(x,-d)}$ are equal in W^* for any $(x,d) \in \mathbb{R}^n \times \mathbb{S}^{n-1}$. In other words, elements of $\mathcal{D}$ can be equivalently viewed as *unoriented* varifolds, i.e. distributions on the product of $\mathbb{R}^n$ and the projective space of $\mathbb{R}^n$, similarly to the framework of [7]. In that particular situation, one obtains an induced distance under the conditions stated in the following proposition which proof is a straightforward adaptation of the one of Proposition 2.2.

Proposition 2.3: *If the kernel defined by ρ is C_0-universal, γ is an even function with $\gamma(1) > 0$ and $\gamma(u) < \gamma(1)$ for all $u \in (-1,1)$, then $\| \cdot \|_{W^*}$ induces a metric on the space $\overset{\circ}{\mathcal{D}}$ modulo the orientation.*

In Section 5 below, we will discuss more thoroughly and illustrate the effects of those kernel properties on the solutions to registration problems for different cases of discrete distributions.

3. Optimal diffeomorphic mapping of varifolds

It is essential to point out that the notion of varifold presented above contains but is also more general than curves and surfaces as it allows to model more complex geometric structures like objects carrying multiple orientation vectors at a given position. In contrast with most previous works on diffeomorphic registration that only involve varifolds as an intermediary representation to compute fidelity terms between shapes, the purpose of this paper to derive a deformation model and registration framework on the space $\mathcal{D}$ itself.

3.1. *Group action*

A first key element is to express the way that deformations 'act' on discrete varifolds. Considering a smooth diffeomorphism $\phi \in \text{Diff}(\mathbb{R}^n)$, we first intend to express how ϕ should transport a Dirac $\delta_{(x,d)}$. There is however not a canonical way to define it as the nature of the underlying data affects the deformation model itself. An important distinction to be made is on the interpretation of direction vectors d, whether they correspond for instance to a unit tangent direction to a curve or a surface in which case d is transported by the Jacobian of ϕ as $D_x\phi(d)/|D_x\phi(d)|$ or rather to a normal direction which instead requires a transport model involving the inverse of the transposed Jacobian i.e. $(D_x\phi)^{-T}(d)/|(D_x\phi)^{-T}(d)|$ (see [22] chap. 10 for more thorough discussion). To keep notations more compact, we will write $D\phi \cdot d$ for a given generic action of $D\phi$ on $\mathbb{R}^n$ on either tangent or normal vector and $\overline{D\phi \cdot d}$ for the corresponding normalized vector in $\mathbb{S}^{n-1}$. That being said, we will also consider two distinct models for the action:

- $\phi_*\delta_{(x,d)} \doteq \delta_{(\phi(x),\overline{D\phi \cdot d})}$ (*normalized action*): this corresponds to transporting the Dirac mass at the new position $\phi(x)$ and transforming the orientation vector as $\overline{D\phi \cdot d}$.
- $\phi_\#\delta_{(x,d)} \doteq |D\phi \cdot d|\delta_{(\phi(x),\overline{D\phi \cdot d})}$ (*pushforward action*): the position and orientation vector are transported as previously but with a reweighting factor equal to the norm of $D\phi \cdot d$.

It is then straightforward to extend both of these definitions by linearity to any discrete varifold in $\mathcal{D}$. In both cases, we obtain a group action of diffeomorphisms on the set of discrete varifolds. However, these actions are clearly not equivalent. The normalized action operates as a pure transport of mass and rotation of the direction vector whereas the pushforward model adds a weight change corresponding to the Jacobian of ϕ along the direction d. This is a necessary term in the situation where $\mu = \mu_S$ is representing a discrete oriented curve or surface. Indeed, one can check, up to discretization errors, that under the pushforward model, we have $\phi_\#\mu_S = \mu_{\phi(S)}$; in other words the action is compatible with the usual deformation of a shape. In the result section below, we will show examples of matching based on those different group action models.

3.2. *Orbits and isotropy groups*

Although we will be focusing on special subgroups of diffeomorphisms in the next section, it will be insightful to study a little more closely the orbits of discrete varifolds under the normalized and pushforward actions of the full group $\mathrm{Diff}(\mathbb{R}^n)$ (or similarly the equivalence classes $\mathcal{D}/\mathrm{Diff}(\mathbb{R}^n)$). Let $\mu \in \mathcal{D}$ which we can write as $\mu = \sum_{i=1}^{N} \sum_{j=1}^{n_i} r_{i,j} \delta_{(x_i,d_{i,j})}$ where the x_i are here assumed to be distinct positions and for each $i = 1,\ldots,N$, the $(d_{i,j})_{j=1,\ldots,n_i}$ are distinct in $\mathbb{S}^{n-1}$. While it is well-known that $\mathrm{Diff}(\mathbb{R}^n)$ acts transitively on the set of point clouds of N points in $\mathbb{R}^n$ (as $n \geq 2$), this may no longer hold when one or several direction vectors are attached to each point position.

In the case of the normalized action, we have $\phi_*\mu = \sum_{i=1}^{N} \sum_{j=1}^{n_i} r_{i,j} \delta_{(\phi(x_i),\overline{D\phi \cdot d_{i,j}})}$. We see that the orbit of μ is then given by:

$$\mathrm{Diff}_*\mu = \left\{ \nu \in \mathcal{D} \ s.t \ \exists(y_i) \in (\mathbb{R}^n)^N, \ y_i \neq y_j \text{ for } i \neq j, \ \exists A_1,\ldots,A_N \in \mathrm{GL}(\mathbb{R}^n), \right.$$

$$\left. \nu = \sum_{i=1}^{N} \sum_{j=1}^{n_i} r_{i,j} \delta_{(y_i,u_{i,j})} \text{ with } u_{i,j} = \frac{A_i d_{i,j}}{|A_i d_{i,j}|} \right\}.$$

This is essentially the set of all discrete varifolds with any set of N distinct positions and for each i, a set of n_i directions obtained by a linear transformation of the $\{d_{i,j}\}_{j=1,\ldots,n_i}$ with weights $r_{i,j}$ unchanged. In particular, this imposes some constraints on the set of 'attainable' direction vectors: clearly, if the number of direction vectors at a given position exceeds the dimension i.e. $n_i \geq n$, this system of vectors cannot be mapped in general to any other system of n_i vectors on the sphere by a single linear map. If we assume that the system of vectors at each position x_i forms a frame, i.e. that for all i, $n_i \leq n$ and the direction vectors $d_{i,j}$ for $j = 1,\ldots,n_i$ are independent, then we see that the orbit of μ is given by the set of all discrete varifolds of the form $\sum_{i=1}^{N} \sum_{j=1}^{n_i} r_{i,j} \delta_{(y_i,u_{i,j})}$ with distinct y_i's and $(u_{i,j})$ in $\mathbb{S}^{n-1}$ such that the $(u_{i,j})_{j=1,\ldots,n_i}$ are independent for all i. In the special case of $n_i = 1$ for all i, that is $\mu \in \overset{\circ}{\mathcal{D}}$, the orbits are then entirely determined by the set of weights r_i which gives the identification of $\overset{\circ}{\mathcal{D}}/\mathrm{Diff}(\mathbb{R}^n)$ with ordered finite sets of positive numbers.

With pushforward action, we have $\phi_\#\mu = \sum_{i=1}^{N} \sum_{j=1}^{n_i} |D\phi \cdot d_{i,j}| r_{i,j} \delta_{(\phi(x_i),\overline{D\phi \cdot d_{i,j}})}$ and the orbit writes:

$$\mathrm{Diff}_\#\mu = \left\{ \nu \in \mathcal{D} \ s.t \ \exists(y_i) \in (\mathbb{R}^n)^N, \ y_i \neq y_j \text{ for } i \neq j, \ \exists A_1,\ldots,A_N \in \mathrm{GL}(\mathbb{R}^n), \right.$$

$$\left. \nu = \sum_{i=1}^{N} \sum_{j=1}^{n_i} |A_i d_{i,j}| r_{i,j} \delta_{(y_i,u_{i,j})} \text{ with } u_{i,j} = \frac{A_i d_{i,j}}{|A_i d_{i,j}|} \right\}.$$

In the general situation, there is again no simple characterization of the orbit.

With the additional assumptions that $n_i \leq n$ and the $(d_{i,j})_{j=1,\ldots,n_i}$ are independent vectors for each i, the orbit of μ is the set of all discrete varifolds of the form $\sum_{i=1}^{N} \sum_{j=1}^{n_i} s_{i,j} \delta_{(y_i, u_{i,j})}$ with any choice of distinct points y_i, direction vectors $(d_{i,j})$ in $\mathbb{S}^{n-1}$ such that the $(d_{i,j})_j$ are independent and weights $s_{i,j} > 0$. In particular, the action of $\mathrm{Diff}(\mathbb{R}^n)$ in the pushforward model is transitive on all subsets of $\overset{\circ}{\mathcal{D}}$ with fixed N, which implies that the equivalence classes of $\overset{\circ}{\mathcal{D}}/\mathrm{Diff}(\mathbb{R}^n)$ in that case are only determined by the number of Diracs in the discrete varifold, as we would expect.

The previous discussion thus shows that for both models and unlike the more standard cases of landmarks or discrete vector fields, the action of diffeomorphisms on discrete varifolds is in general not transitive. It is therefore necessary to formulate the registration problems in their inexact form by introducing fidelity terms like the kernel metrics introduced in Section 2.

An alternative viewpoint is to identify the previous orbits as quotients of $\mathrm{Diff}(\mathbb{R}^n)$ by their isotropy subgroups, which allows to draw some parallel with the model of jets of [14]. To simplify the presentation, let us consider $\mu = \sum_{i=1}^{N} r_i \delta_{(x_i, d_i)}$ in $\overset{\circ}{\mathcal{D}}$ (thus $x_i \neq x_j$ for $i \neq j$). Writing G_μ^* and $G_\mu^\#$ the isotropy subgroup of μ for the normalized and pushforward actions, we have the identifications $\mathrm{Diff}(\mathbb{R}^n)/G_\mu^* \simeq \mathrm{Diff}_*\mu$ and $\mathrm{Diff}(\mathbb{R}^n)/G_\mu^\# \simeq \mathrm{Diff}_\#\mu$. Now, we can describe those isotropy subgroups as follows. Denoting $\mathcal{S}_N$ the permutation group of $\{1,\ldots,N\}$, we have

$$G_\mu^* = \left\{ \phi \in \mathrm{Diff}(\mathbb{R}^n) \mid \exists \pi \in \mathcal{S}_N, \, \forall i = 1,\ldots,N, \, \phi(x_i) = x_{\pi(i)} \text{ and } \frac{d_{x_i}\phi(d_i)}{|d_{x_i}\phi(d_i)|} = d_{\pi(i)} \right\}$$

and

$$G_\mu^\# = \left\{ \phi \in \mathrm{Diff}(\mathbb{R}^n) \mid \exists \pi \in \mathcal{S}_N, \, \forall i = 1,\ldots,N, \, \phi(x_i) = x_{\pi(i)} \text{ and } r_i d_{x_i}\phi(d_i) = r_{\pi(i)} d_{\pi(i)} \right\}.$$

Interestingly, the isotropy group for the space of first-order jets on the landmark set $z = \{x_i\}_{i=1,\ldots,N}$ as derived in [14] is given by

$$G_z^{(1)} = \{ \phi \in \mathrm{Diff}(\mathbb{R}^n) \mid \forall i = 1,\ldots,N, \, \phi(x_i) = x_i \text{ and } d_{x_i}\phi = \mathrm{Id} \}.$$

We can notice that $G_z^{(1)}$ is actually a subgroup of both G_μ^* and $G_\mu^\#$, which shows again that the deformation models on discrete varifolds considered here come with additional invariances, in particular the invariance to permutation of the Dirac varifolds in μ, in contrast with the labelled particles of [20, 14].

3.3. *Optimal control problem*

With the definitions and notations of the previous sections, we can now introduce the mathematical formulation of the diffeomorphic registration of discrete varifolds. As mentioned in the introduction, we will rely on the LDDMM model for generating diffeomorphisms although other transformation spaces and models could be taken as well. In short, we consider a space of time dependent velocity fields $v \in L^2([0,1],V)$ such that for all $t \in [0,1]$, v_t belongs to a certain RKHS V of vector fields on $\mathbb{R}^n$. We will write $K : \mathbb{R}^n \times \mathbb{R}^n \to \mathbb{R}^n$ the vector-valued reproducing kernel of V. From v, one obtains the flow mapping ϕ_t^v at each time t as the integral of the differential equation $\partial_t \phi_t^v = v_t \circ \phi_t^v$ with $\phi_0 = \mathrm{Id}$. The deformation group G_V is then defined as the set of all flow maps ϕ_1^v for all velocity fields $v \in L^2([0,1],V)$. With the adequate assumptions on the kernel of V, this is a subgroup of $\mathrm{Diff}(\mathbb{R}^n)$ and it is naturally equipped with a *right-invariant metric* given on V by $\int_0^1 \|v_t\|_V^2 dt$. This right-invariant metric can be then shown to descend to a metric on the orbit spaces of discrete varifolds for the action of G_V, which can be described in a similar way as in the previous section. We refer to [22] for a detailed exposition of the LDDMM framework.

Now, consider two discrete varifolds $\mu_0 = \sum_{i=1}^P r_{i,0}\delta_{(x_{i,0},d_{i,0})}$ (template) and $\tilde{\mu} = \sum_{j=1}^Q \tilde{r}_j \delta_{(\tilde{x}_j,\tilde{d}_j)}$ (target). As μ_0 and $\tilde{\mu}$ may very well not belong to the same orbit, we introduce the registration problem in its inexact formulation which consists in the variational problem:

$$\mathrm{argmin}_{v \in L^2([0,1],V)} \left\{ E(v) = \int_0^1 \|v_t\|_V^2 dt + \lambda \|\mu(1) - \tilde{\mu}\|_{W^*}^2 \right\} \tag{3.1}$$

subject to either $\mu(t) \doteq (\phi_t^v)_* \mu_0$ in the normalized action scenario or $\mu(t) \doteq (\phi_t^v)_\# \mu_0$ for the pushforward model, and λ being a weight parameter between the regularization and fidelity terms in the energy. This is easily interpreted as an optimal control problem in which the state variable is the transported varifold $\mu(t)$, the control is the velocity field v and the cost functional is the sum of the standard LDDMM metric on the deformation and a discrepancy term between $\mu(1)$ and the target given by a varifold kernel metric as in (2.1). Those optimal control problems are well-posed in the following sense:

Proposition 3.1: *If V is continuously embedded in the space $C_0^2(\mathbb{R}^n, \mathbb{R}^n)$, or equivalently if K is of class C^2 with all derivatives up to order 2 vanishing at infinity, then there exists a global minimum to the problem* (3.1).

Proof: The result follows from an argument similar to that of the existence of minimizers in usual LDDMM registration problems. If (v^n) is a minimizing sequence in $L^2([0,1],V)$ then thanks the first term of E, we may assume that (v^n)

is bounded in $L^2([0,1],V)$ and therefore that, up to extracting a subsequence, $v^n \rightharpoonup v^*$ weakly in $L^2([0,1],V)$. It then follows from the results of [22] (Chapter 8.2) that the sequence of diffeomorphisms $(\phi_1^{v^n})$ and their first-order differentials $(d\phi_1^{v^n})$ converge uniformly on every compact respectively to $\phi_1^{v^*}$ and $d\phi_1^{v^*}$. In particular, for all $i = 1, \ldots, P$, $\phi_1^{v^n}(x_i) \to \phi_1^{v^*}(x_i)$ and $d\phi_1^{v^n}(x_i) \to d\phi_1^{v^*}(x_i)$. Then, from the expressions of the group actions and the metric (2.1), we obtain that either $\|(\phi_1^{v^n})_*\mu_0 - \tilde{\mu}\|_{W^*}^2 \xrightarrow[n\to\infty]{} \|(\phi_1^{v^*})_*\mu_0 - \tilde{\mu}\|_{W^*}^2$ or $\|(\phi_1^{v^n})_\#\mu_0 - \tilde{\mu}\|_{W^*}^2 \xrightarrow[n\to\infty]{}$ $\|(\phi_1^{v^*})_\#\mu_0 - \tilde{\mu}\|_{W^*}^2$. Finally, using the weak lower semicontinuity of the norm in $L^2([0,1],V)$, it gives in both cases:

$$E(v^*) \le \liminf_{n\to\infty} E(v^n)$$

and consequently v^* is a global minimizer of E. $\square$

3.4. Hamiltonian dynamics

By fixing the final time condition $\mu(1)$ and minimizing $\int_0^1 \|v_t\|_V^2 dt$ with those boundary constraints, the resulting path $t \mapsto \mu(t)$ corresponds to a *geodesic* in $\mathcal{D}$ for the metric induced by the metric on the deformation group. We can further characterize those geodesics as solutions of a Hamiltonian system. For that purpose, we follow the general setting developed in [2] for similar optimal control problems.

In our situation, we can describe the state $\mu(t)$ as a set of P particles each given by the triplet $(x_i(t), d_i(t), r_i(t)) \in \mathbb{R}^n \times \mathbb{S}^{n-1} \times \mathbb{R}_+^*$ representing its position, orientation vector and weight. From Section 3.1, we have that $x_i(t) = \phi_t^v(x_{i,0})$, $d_i(t) = \overline{D\phi_t^v \cdot d_{i,0}}$ and $r_i(t) = r_{i,0}$ for the normalized action and $r_i(t) = |D\phi_t^v \cdot d_{i,0}|r_{i,0}$ in the pushforward case. Differentiating with respect to t, the state evolution may be alternatively described by the set of ODEs

$$\begin{cases} \dot{x}_i(t) = v_t(x_i(t)) \\ \dot{d}_i(t) = P_{d_i(t)^\perp}(Dv_t \cdot d_i(t)) \\ \dot{r}_i(t) = \begin{cases} 0 & \text{(normalized)} \\ \langle d_i(t), Dv_t \cdot d_i(t) \rangle r_i(t) & \text{(pushforward)} \end{cases} \end{cases}$$

where $P_{d_i(t)^\perp}$ denotes the orthogonal projection on the subspace orthogonal to $d_i(t)$, $Dv_t \cdot d_i(t)$ corresponds to the infinitesimal variation of the action of $D\phi$ on vectors of $\mathbb{R}^n$ introduced in Section 3.1: it is given specifically by $Dv_t \cdot d_i(t) = D_{x_i(t)}v_t(d_i(t))$ in the tangent case and $Dv_t \cdot d_i(t) = -(D_{x_i(t)}v_t)^T(d_i(t))$ in the normal case. Note that other choices of transformation of the weights could be treated quite similarly by modifying accordingly the last equation in the previous system. In what follows, we detail the derivations of the optimality equations in the case of tangent direction vectors for both normalized and pushforward group action models, the situation of normal vectors being easily tackled in similar fashion.

3.4.1. *Normalized action*

In the case of normalized action, $\{r_i(t)\}_{i=1}^P$ are time independent as previous discussed. So we can choose the state variable of the optimal control problem to be $q := \{(x_i, d_i),\ i = 1, \ldots, P\} \in \mathbb{R}^{2dP}$ with the infinitesimal action

$$\xi_q v = \left\{ \left(v(x_i), P_{d_i^\perp}(D_{x_i} v(d_i)) \right),\ i = 1, \ldots, P \right\}$$

and introduce the Hamiltonian

$$
\begin{aligned}
H(p, q, v) &= (p | \xi_q v) - \frac{1}{2} \|v\|_V^2 \\
&= \sum_{i=1}^N \langle p_i^{(1)}, v(x_i) \rangle + \langle P_{d_i^\perp}(p_i^{(2)}), D_{x_i} v(d_i) \rangle - \frac{1}{2} \|v\|_V^2,
\end{aligned}
$$

where

$$p = \left\{ \left(p_i^{(1)}, p_i^{(2)} \right),\ i = 1, \ldots, P \right\} \in \mathbb{R}^{2dP}$$

is the adjoint variable of state q. We call $p_i^{(1)}$ the *spatial momentum* and $p_i^{(2)}$ the *directional momentum*. From Pontryagin's maximum principle, the Hamiltonian dynamics is given by the forward system of equations

$$
\begin{cases}
\dot{x}_i(t) = v_t(x_i(t)) \\
\dot{d}_i(t) = P_{d_i(t)^\perp}(D_{x_i(t)} v_t(d_i(t))) \\
\dot{p}_i^{(1)}(t) = -(D_{x_i(t)} v_t)^T p_i^{(1)}(t) \\
\qquad\quad -(D_{x_i(t)}^{(2)} v_t(\cdot, d_i(t)))^T \left(P_{d_i(t)^\perp}(p_i^{(2)}(t)) \right) \\
\dot{p}_i^{(2)}(t) = -(D_{x_i(t)} v_t)^T P_{d_i(t)^\perp}(p_i^{(2)}(t)) \\
\qquad\quad + \langle d_i(t), p_i^{(2)}(t) \rangle D_{x_i(t)} v_t(d_i(t)) \\
\qquad\quad + \langle d_i(t), D_{x_i(t)} v_t(d_i(t)) \rangle p_i^{(2)}(t)
\end{cases}
\tag{3.2}
$$

and optimal vector fields v satisfy

$$
\begin{aligned}
\langle v_t, h \rangle_V &= \left(p(t), \xi_{q(t)} h \right) \\
&= \sum_{i=1}^P \langle p_i^{(1)}, h(x_i) \rangle + \langle P_{d_i(t)^\perp}(p_i^{(2)}(t)), D_{x_i(t)} h(d_i(t)) \rangle,
\end{aligned}
$$

for any $h \in V$ and $t \in [0,1]$. The reproducing property and reproducing property for the derivatives in a vector RKHS give [20] that $\forall x \in \mathbb{R}^n$, $z \in \mathbb{R}^n, v \in V$ and multi-index α,

$$\langle K(x,\cdot)z, v \rangle_V = (z \otimes \delta_x | v)$$
$$\langle D_1^\alpha K(x,\cdot), v \rangle_V = z^T D^\alpha v(x).$$

With the above properties, we obtain the following expression of v

$$v_t(\cdot) = \sum_{k=1}^{P} K(x_k(t),\cdot) p_k^{(1)}(t)$$
$$+ D_1 K(x_k(t),\cdot) \left(d_k(t), P_{d_k(t)^\perp}(p_k^{(2)}(t)) \right), \quad (3.3)$$

where we use the shortcut notation $D_1 K(x,\cdot)(u_1, u_2)$ for the vector $D_1(K(x,\cdot)u_2)(u_1)$. In Figure 2, we show an example of geodesic and resulting deformation for a single Dirac varifold, which is obtained as the solution of (3.2) with the initial momenta shown in the figure. It illustrates the combined effects of the spatial momentum which displaces the position of the Dirac and of the directional momentum that generates a local rotation of the direction vector.

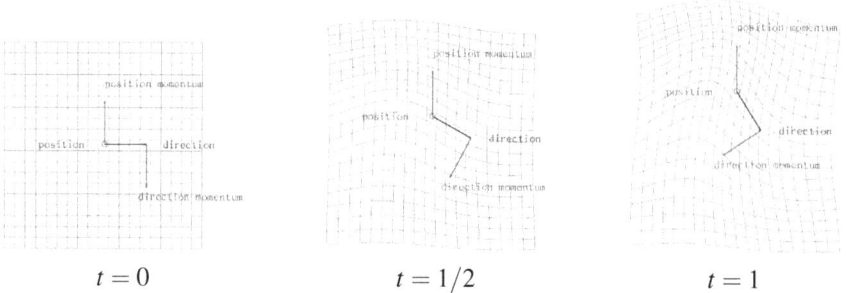

$t = 0$ $t = 1/2$ $t = 1$

Fig. 2. Example of geodesic for a single Dirac in the normalized action case.

3.4.2. Pushforward action

As in the previous section, we set the state variable $q := \{(x_i, d_i, r_i), i = 1, \ldots, P\} \in (\mathbb{R}^n \times \mathbb{S}^{n-1} \times \mathbb{R}_+^*)^P$, the infinitesimal action

$$\xi_q v = \left\{ \left(v(x_i), P_{d_i^\perp}(d_{x_i} v(d_i)), r_i \langle d_i, d_{x_i} v(d_i) \rangle \right), i = 1, \ldots, P \right\}$$

and the Hamiltonian

$$H(p,q,v) = \sum_{i=1}^{P} \left\langle p_i^{(1)}, v(x_i) \right\rangle + \left\langle P_{d_i^{\perp}}(p_i^{(2)}), d_{x_i}v(d_i) \right\rangle + p_i^{(3)} r_i \left\langle d_i, d_{x_i}v(d_i) \right\rangle$$
$$- \frac{1}{2}\|v\|_V^2, \tag{3.4}$$

where $p = \left\{ \left(p_i^{(1)}, p_i^{(2)}, p_i^{(3)} \right), \, i = 1, \ldots, P \right\} \in \mathbb{R}^{(2d+1)P}$. Applying again Pontryagin's maximum principle, we obtain the forward system

$$\begin{cases} \dot{x}_i &= v_t(x_i) \\ \dot{d}_i &= P_{d_i^{\perp}}(d_{x_i}v(d_i)) \\ \dot{r}_i &= r_i \left\langle d_i, d_{x_i}v_t(d_i) \right\rangle \\ \dot{p}_i^{(1)} &= -(d_{x_i(t)}v_t)^T p_i^{(1)} - (d_{x_i(t)}^{(2)}v(\cdot, d_i))^T (P_{d_i^{\perp}}(p_i^{(2)})) - p_i^{(3)} r_i d_{x_i}v(\cdot, d_i)^T d_i \\ \dot{p}_i^{(2)} &= -d_{x_i(t)}v_t^T(p_i^{(2)}) + \left(\left\langle d_i, p_i^{(2)} \right\rangle - r_i p_i^{(3)} \right) \left[d_{x_i}v_t + d_{x_i}v_t^T \right] (d_i) \\ & \quad + \left\langle d_i, d_{x_i}v_t(d_i) \right\rangle p_i^{(2)} \\ \dot{p}_i^{(3)} &= -p_i^{(3)} \left\langle d_i, d_{x_i}v_t(d_i) \right\rangle \end{cases} \tag{3.5}$$

with optimal vector field of the form

$$v_t(x) = \sum_{k=1}^{P} K(x_k, x) p_k^{(1)} + D_1 K(x_k, x) \left(d_k, P_{d_k^{\perp}}(p_k^{(2)}) + p_k^{(3)} r_k d_k \right). \tag{3.6}$$

From the forward equations (3.5), we see that $\frac{d}{dt}\langle d_i(t), d_i(t) \rangle = 0$ and $\frac{d}{dt} r_i(t) p_i^{(3)}(t) = 0$, hence $\|d_i(t)\|$ and $r_i(t) p_i^{(3)}(t)$ are constant along geodesic paths. Similarly to the normalized action case, we can use those conservation properties to reduce the number of state and dual variables as follows.

Let the new state variable be $q = \{(x_i, u_i), \, i = 1, \ldots, P\}$ and the Hamiltonian

$$H(p,q,v) = \sum_{i=1}^{P} \langle p_i^{(1)}, v(x_i) \rangle + \langle p_i^{(2)}, d_{x_i}v(u_i) \rangle - \frac{1}{2}\|v\|_V^2. \tag{3.7}$$

The forward equations and optimal vector field v derived from this Hamiltonian are

$$\begin{cases} \dot{x}_i(t) = v_t(x_i(t)) \\ \dot{d}_i(t) = d_{x_i(t)}v_t(u_i(t)) \\ \dot{p}_i^{(1)}(t) = -(d_{x_i(t)}v_t)^T p_i^{(1)} - (d_{x_i(t)}^{(2)}v(\cdot, u_i))^T p_i^{(2)} \\ \dot{p}_i^{(2)}(t) = -(d_{x_i(t)}v_t)^T p_i^{(2)}(t) \end{cases} \tag{3.8}$$

and

$$v_t(x) = \sum_{k=1}^{P} K(x_k,x)p_k^{(1)}(t) + D_1 K(x_k,x)(u_k(t), p_k^{(2)}(t)). \qquad (3.9)$$

Then this new system is rigorously equivalent to the original one in the following sense:

Proposition 3.2: *Any solution of (3.8) + (3.9) is such that* $\left(x_i(t), \overline{u_i(t)}, |u_i(t)|\right)$ *is a solution of (3.5) + (3.6). Conversely, any solution* $(x_i(t), d_i(t), r_i(t))$ *of (3.5) + (3.6) with initial conditions satisfying* $\left\langle p_i^{(2)}(0), u_i(0) \right\rangle = r_i(0)p_i^{(3)}(0)$ *gives the solution* $(x_i(t), r_i(t)d_i(t))$ *to (3.8) + (3.9).*

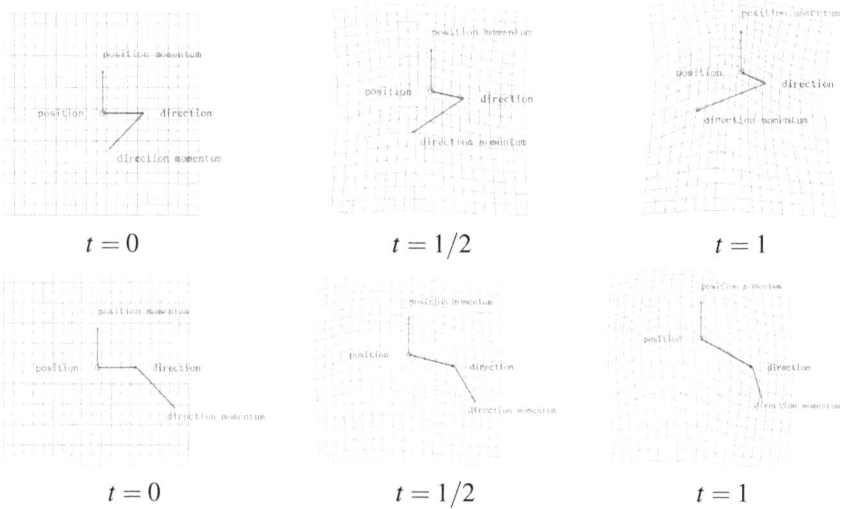

$t=0$ $t=1/2$ $t=1$

$t=0$ $t=1/2$ $t=1$

Fig. 3.   Examples of geodesics in the pushforward action case.

    The proof is given in Appendix. Note that these equations can be also obtained in a more particular case as the geodesic equations on the tangent bundle of the space of landmarks, as derived for instance in [1] (Section 3.5). In what follows, we will thus replace the system (3.5) by (3.8).

Remark 3.3: We point out that there are other conserved quantities in the previous system. In particular, it's easy to see that $\left\langle p_i^{(2)}, d_i \right\rangle$ is constant along geodesics since

$$\frac{d}{dt}\left\langle p_i^{(2)}, d_i \right\rangle = -\left\langle (d_{x_i}v_t)^T p_i^{(2)}, d_i \right\rangle + \left\langle p_i^{(i)}, d_{x_i}v_t(d_i) \right\rangle = 0.$$

Figure 3 shows two geodesic trajectories of a single Dirac varifold for different initial momenta. In particular, we can again observe the effect of the directional momentum $p^{(2)}$ on the dynamics and resulting deformations. In addition to similar rotation effects as in the normalized action case, local contraction or expansion can be generated as well, depending precisely on the angle $\langle p_i^{(2)}, d_i \rangle$.

4. Registration algorithm and implementation

We now turn to the issue of numerically solving the optimization problem (3.1). We will follow the commonly used method for such problems called *geodesic shooting* (cf [21]). Indeed, from the developments of Section 3.4, we see that optimizing (3.1) with respect to vector fields v can be done equivalently by restricting to geodesics and thus by optimizing over the initial momenta variables $p_0^{(1)}$ and $p_0^{(2)}$ that completely parametrize those geodesics through the Hamiltonian equations.

4.1. *Computation of E*

Let a template and target discrete varifold be given as in Section 3.3. As mentioned above, we can rewrite the energy E as a function of the initial momenta that we will denote $p_0^{(1)} = (p_i^{(1)}(0))$ and $p_0^{(2)} = (p_i^{(2)}(0))$:

$$E(p_0^{(1)}, p_0^{(2)}) = H_r(p_0, q_0) + \lambda \underbrace{\|\mu(1) - \tilde{\mu}\|_{W^*}^2}_{:=g(q(1))} \tag{4.1}$$

where q_0 is the initial state, $\mu(1)$ is the varifold corresponding to the final time state $q(1)$ with $g(q(1))$ the resulting fidelity term between $\mu(1)$ and the target varifold, and H_r is the *reduced Hamiltonian* $H_r(p,q) = H(p,q,v)$ for the optimal v given by (3.3) or (3.9) (note that $H_r(p(t), q(t))$ is conserved along solutions of the Hamiltonian systems thus giving the above expression of the energy).

The expression of H_r as well as the resulting reduced Hamiltonian equations can be obtained in all generality by plugging the expression of v in the equations of Section 3.3. In our implementation, we actually restrict to the more particular case of radial scalar kernels for the vector fields in V, i.e. we assume that $K(x,y) = h(|x-y|^2)I_n$. Then the reduced Hamiltonian for the normalized case becomes:

$$H_r(p,q) = \frac{1}{2} \langle \overline{K}_q p, p \rangle, \tag{4.2}$$

where $\overline{K}_q$ is a symmetric positive definite matrix which is defined as follows. Let

$$H = (H)_{ik} = (h_{ki})$$
$$\overline{A} = (\overline{A})_{ik} = 2\dot{h}_{ki}\langle x_k - x_i, d_k\rangle$$
$$\overline{B} = (\overline{B})_{ik} = -\left[4\ddot{h}_{ik}\langle x_k - x_i, d_i\rangle\langle x_k - x_i, d_k\rangle + 2\dot{h}_{ki}\langle d_i, d_k\rangle\right]$$

with h_{ki} being a shortcut for $h(|x_i - x_j|^2)$ and

$$P_{d^\perp}(\cdot) = \begin{pmatrix} I_d - d_1 \cdot d_1^T & & 0 \\ & \ddots & \\ 0 & & I_d - d_N \cdot d_N^T \end{pmatrix}.$$

Then we define

$$\overline{K}_q := \begin{pmatrix} I_{Pd} & 0 \\ 0 & P_{d^\perp} \end{pmatrix}^T \begin{pmatrix} H \otimes I_d & \overline{A} \otimes I_d \\ (\overline{A} \otimes I_d)^T & \overline{B} \otimes I_d \end{pmatrix} \begin{pmatrix} I_{Pd} & 0 \\ 0 & P_{d^\perp} \end{pmatrix}$$

where $\otimes$ denotes the Kronecker product. For the pushforward action case, we define H, A and B as in normalized action case with d_i and d_k replaced by u_i and u_k, then

$$H_r(p,q) = \frac{1}{2}\langle K_q p, p\rangle, \tag{4.3}$$

where

$$K_q := \begin{pmatrix} H \otimes I_d & A \otimes I_d \\ (A \otimes I_d)^T & B \otimes I_d \end{pmatrix}.$$

This gives us explicitly the first term of the energy in (4.1).

Now, the time evolution of q and p can be also rewritten equivalently in reduced Hamiltonian form, which expressions are given in full for radial scalar kernels in the Appendix. We numerically integrate those differential systems using an RK4 scheme, which we experienced to be better-adapted to these systems than the simpler Euler midpoint integrator used in [16]. Then, given initial momenta $p_0^{(1)}$ and $p_0^{(2)}$, integrating those equations forward in time produces the final state $q(1)$ and its corresponding varifold $\mu(1)$. It is then straightforward to evaluate the second term in (4.1) through the expression of the varifold norm (2.1); in the pushforward case one only needs to apply the additional intermediate operation of converting state $q(1) = (x_i(1), u_i(1))$ into $(x_i(1), \overline{u}_i(1), |u_i|(1))$. We will discuss different choices of kernels for the varifold metric in the result section.

4.2. *Computation of the gradient of E*

The second element we need is the gradient of the energy with respect to the momenta. The first term being directly a function of p_0, it can be differentiated easily and gives the following gradient:

$$\nabla_{p_0} H_r(p_0, q_0) = \overline{K}_q p_0 \quad \text{(normalized)}$$
$$\nabla_{p_0} H_r(p_0, q_0) = K_q p_0 \quad \text{(pushforward)}.$$

The fidelity term $g(q(1))$ in (4.1), however, is a function of the final state $q(1)$ which is in turn a function of the momenta through the Hamiltonian system of equations. The computation of the gradient is therefore more involved due to the complicated dependency of $q(1)$ in p_0. The standard approach for optimal control problems of this form (cf for example [21] or [2]) is to introduce the *adjoint Hamiltonian system*:

$$\dot{Z}(t) = d(\partial_p H_r, \partial_q H_r)^T Z(t)$$

with $Z = (\tilde{q}, \tilde{p})^T$ the vector of the adjoint variables. Then, as detailed in [2], the gradient of $g(q(1))$ with respect to p_0 is given by $\tilde{p}(0)$ where $(\tilde{q}(t), \tilde{p}(t))^T$ is the solution of the adjoint system integrated backward in time with $\tilde{q}(1) = \nabla g(q(1))$ and $\tilde{p}(1) = 0$.

For the particular Hamiltonian equations considered here, the adjoint system is tedious to derive and to implement. We simply avoid that by approximating the differentials appearing in the adjoint system by finite difference of the forward Hamiltonian equations, following the suggestion of [2] (Section 4.1) which we refer to for details. Note that another possibility would be to take advantage of automatic differentiation methods, as used recently for some LDDMM registration problems by the authors of [17].

Lastly, the end time condition $\nabla g(q(1))$ in the previous adjoint system is computed by direct differentiation of the varifold norm (2.1) with respect to the final state variables. This is actually more direct than in previous works like [7, 16] where the gradients are computed with respect to the positions of vertices of the underlying mesh. Here, we have specifically, for the normalized model:

$$\partial_{x_i} g(q(1)) = 2 \sum_{j=1}^{P} 2 r_i r_j \rho'(|x_i(1) - x_j(1)|^2) \gamma(\langle d_i(1), d_j(1) \rangle).(x_i(1) - x_j(1)) \; -2 \ldots$$

$$\partial_{d_i} g(q(1)) = 2 \sum_{j=1}^{P} r_i r_j \rho(|x_i(1) - x_j(1)|^2) \gamma'(\langle d_i(1), d_j(1) \rangle).d_j(1) \; -2 \ldots$$

where the ... denote a similar term for the differential of the cross inner product $\langle \mu(1), \tilde{\mu} \rangle_{W^*}$. In the pushforward case with state variables $(x_i(1), u_i(1))$, we first compute $d_i(1) = u_i(1)/|u_i(1)|$ and $r_i(1) = |u_i(1)|$ and obtain $\partial_{x_i} g(q(1))$ with the same expression as above while $\partial_{u_i} g(q(1))$ is given by a simple chain rule.

Finally, with the above notations, the gradient of E writes:

$$\nabla_{p_0} E = \overline{K}_q p_0 + \lambda \tilde{p}(0) \tag{4.4}$$

respectively $\nabla_{p_0} E = K_q p_0 + \lambda \tilde{p}(0)$ in the pushforward case.

4.3. *Gradient descent algorithm*

The solution to the minimization of (4.1) is then computed by gradient descent on $p_0 = \left(p_{0,i}^{(1)}, p_{0,i}^{(2)} \right)_{i=1,\dots,P}$. Note that this is a non-convex optimization problem. Until convergence, each iteration consists of the following steps:

(1) Given the current estimate of p_0, integrate the Hamiltonian equations forward in time to obtain $q(1)$.

(2) Compute the gradient $\nabla g(q(1))$.

(3) Integrate the adjoint Hamiltonian system backward in time to obtain $\nabla_{p_0} E$.

(4) Update p_0: we use two separate update steps for the spatial and directional momentum which are selected, at each iteration, using a rough space search approach leading to the lowest value of E.

5. Results

We now present a few results of registration using the previous algorithm on simple and synthetic examples. Our implementation equally supports objects in 2D or 3D, we will however focus on examples in $\mathbb{R}^2$ here simply to allow for an easier visualization and interpretation of the results.

5.1. *Curve registration*

We begin with a toy example of standard curve matching to compare the result and performance of our discrete varifold LDDMM registration algorithm with the state-of-the-art LDDMM approach for curves such as the implementations of [10, 16]. The former methods share a very similar formulation to (3.1) and also make use of varifold metrics as fidelity terms, the essential difference being that the state of the optimal control problem is there the set of vertices of the deformed template curve which is only converted to a varifold for the evaluation of the fidelity term at each iteration. But the dynamics of geodesics still correspond to usual point set deformation under the LDDMM model.

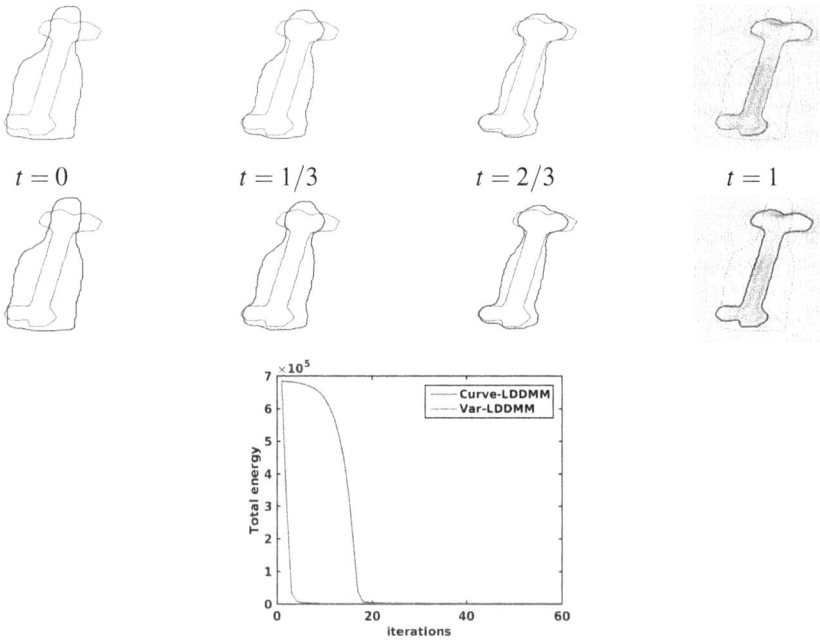

Fig. 4. Curve registration using point-mesh LDDMM (1st row) and our proposed discrete varifold LDDMM (2nd row). On the last row is shown the evolution of the total energy across the iterations for both algorithms.

We consider here the pushforward model for the action of diffeomorphisms on discrete varifolds that we have seen is compatible with the action of diffeomorphisms on curves. In this case, the two formulation and optimization problems for curve registration are theoretically equivalent up to discretization precision. We verify it with the example of Figure 4 for which both algorithms are applied with the same deformation kernel, varifold metric and optimization scheme. Note that in our approach, template and target curves are first (and only once at the beginning) converted to their discrete varifold representations as explained in Section 2.

As we can see, the resulting geodesics and deformations are consistent between the two methods. This is also corroborated by the very similar values of the energy at convergence. Interestingly however, although each iteration in our model is arguably more expansive numerically compared to standard curve-LDDMM due to the increased complexity of the Hamiltonian equations, the algorithm converges in a significantly lesser number of iterations. Whether this observation generalizes to other examples or other optimization methods will obviously require more careful examination in future work.

5.2. *Registration of directional sets*

We now turn to examples that are more specific to the framework of discrete varifolds.

Choice of the varifold metric

First, we examine more closely the effect of the metric $\| \cdot \|_{W^*}$ on the registration of discrete varifolds. The framework we propose can indeed support many choices for the kernel functions ρ and γ that define fidelity metrics $\| \cdot \|_{W^*}$ with possibly very different properties. This has been already analyzed quite extensively in [16] but only in the situation where varifolds associated to a curve or a surface. We consider here the same examples of kernels and briefly discuss what are the specific effects to expect when matching more general varifolds in $\mathcal{D}$ which may involve several orientation vectors at a given position.

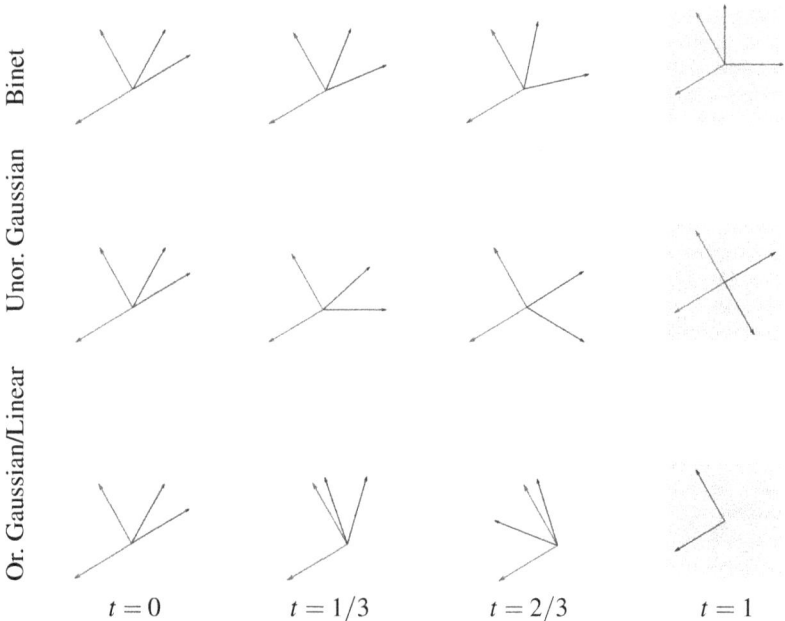

Fig. 5. (Color online) Matching of pairs of Dirac varifolds (template is in blue and target in red) under the normalized action with different choices of kernels: Binet on the first row, unoriented Gaussian ($\sigma_s = 1$) on the second and oriented Gaussian ($\sigma_s = 2$) on the last one. The linear kernel leads to the same result as the former in that particular case.

The results of Propositions 2.2 and 2.3 hold under the assumption that the kernel defined by ρ is a C_0-universal kernel on $\mathbb{R}^n$, which restricts the possible choices to a few known classes (cf [6] for a thorough analysis). Here, we will focus on the class of Gaussian kernels given by $\rho(|x-x'|^2) = e^{-\frac{|x-x'|^2}{\sigma^2}}$ with a width parameter $\sigma > 0$ that essentially provides a notion of spatial scale sensitivity to the metric, and which must be adapted to the intrinsic sizes of shapes in each example.

In combination with ρ, as in [16], we introduce the following four kernels on $\mathbb{S}^{n-1}$:

- $\gamma(\langle d, d' \rangle) = \langle d, d' \rangle$ (linear kernel): this choice is related to the particular subclass of currents [10]. In that case, the resulting $\| \cdot \|_{W^*}$ is clearly only a pseudo-metric on $\mathcal{D}$ since the linearity implies that in W^*: $\delta_{(x,-d)} = -\delta_{(x,d)}$ and for any $d_1 \neq -d_2$, $\delta_{(x,d_1)} + \delta_{(x,d_2)} = |d_1 + d_2| \delta_{\left(x, \frac{d_1+d_2}{|d_1+d_2|}\right)}$.
  However, we still obtain a metric on the subspace $\overset{\circ}{\mathcal{D}}$ thanks to Proposition 2.2.

- $\gamma(\langle d, d' \rangle) = \langle d, d' \rangle^2$ (Binet kernel): γ being an even function, as discussed in Section 2, the resulting metric on W^* is invariant to the orientation of direction vectors. According to Proposition 2.3, we then have a distance on $\overset{\circ}{\mathcal{D}}$ modulo the orientation. Note however that with this particular choice, one does not obtain a metric (but only a pseudo-metric) on $\mathcal{D}$ modulo the orientation, as we will illustrate in the examples below.

- $\gamma(\langle d, d' \rangle) = e^{-\frac{2}{\sigma_S^2}(1 - \langle d, d' \rangle^2)}$ (unoriented Gaussian kernel): this is another example of orientation-invariant kernel considered in [7] corresponding to a particular construction of Gaussian kernels on the projective space. In contrast with Binet kernel, it does induce a metric on $\mathcal{D}$ modulo orientation.

- $\gamma(\langle d, d' \rangle) = e^{-\frac{2}{\sigma_S^2}(1 - \langle d, d' \rangle)}$ (oriented Gaussian kernel): this kernel is the restriction of the standard Gaussian kernel on $\mathbb{R}^n$ to the sphere $\mathbb{S}^{n-1}$. As such, it can be shown to be C_0-universal on $\mathbb{S}^{n-1}$ and thus, from Proposition 2.1, lead to a metric on the entire space $\mathcal{D}$.

We illustrate the aforementioned properties on a very simple registration example between pairs of Dirac varifolds located at the same position x i.e. $\delta_{(x,d_1)} + \delta_{(x,d_2)}$ and $\delta_{(x,d_1')} + \delta_{(x,d_2')}$. In Figure 5, the template and target pairs of Diracs are matched based on the normalized action model. The estimated matching and deformations clearly differ with the choice of kernel but each of these result is in fact perfectly consistent with the different invariances of those kernels. Indeed the two Diracs are exactly matched to the target using the oriented

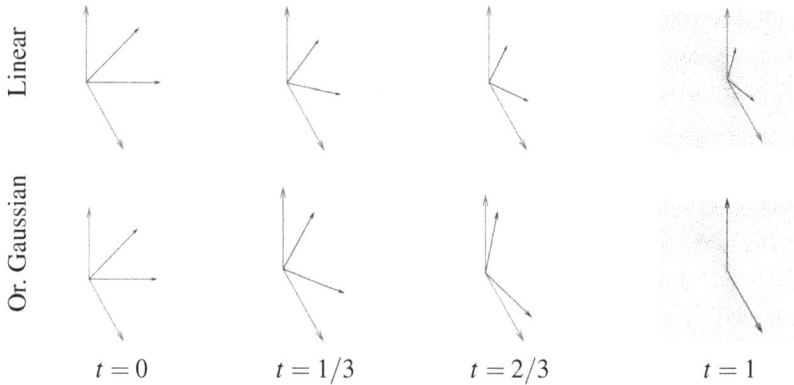

Fig. 6.   Registration of pairs of Dirac varifolds with the pushforward model for both the linear and oriented Gaussian kernel.

Gaussian kernel since $\|\cdot\|_{W^*}$ is in that case a metric on the entire space $\mathcal{D}$. They are however matched to the opposite vectors with the unoriented Gaussian kernel which is indeed insensitive to orientation. In the case of Binet kernel, in addition to orientation-invariance, there exists other pairs of Diracs which are distinct in $\mathcal{D}$ but coincide in W^*. For example, it can be easily verified that all discrete varifolds of the form $\delta_{(x,d_1)} + \delta_{(x,d_2)}$ with orthogonal vectors d_1 and d_2 are equal in W^*, which is reflected by the result in Figure 5.

We emphasize the difference of behavior between linear and oriented Gaussian kernels with the example of Figure 6 associated this time to the pushforward action model. The result shown in the first row is a consequence of the fact that fidelity terms derived from the linear kernel only constrains the sums $d_1 + d_2$ and $d_1' + d_2'$ to match.

Multi-directional varifold matching

Finally, Figure 7 shows an example of matching on more general discrete varifolds that involve varying number of directions at different spatial locations. This is computed with the normalized action using an oriented Gaussian kernel for the fidelity term. Although purely synthetic, it illustrates the potentialities of the proposed approach to register data with complex directional patterns.

5.3. Contrast-invariant image registration

A last possible application worth mentioning is the registration of images with varying contrast. Indeed, an image I modulo all contrast changes is equivalently represented by its unit gradient vector field $\frac{\nabla I}{|\nabla I|}$. Note that this may in fact be only

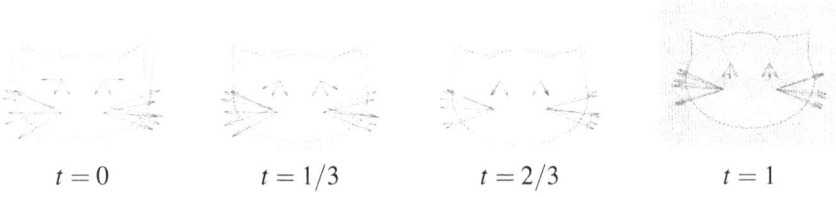

| $t = 0$ | $t = 1/3$ | $t = 2/3$ | $t = 1$ |

Fig. 7. Registration of multi-directional sets. The lengths of vectors correspond to the weights of the Dirac varifolds.

defined at isolated pixels in the image, specifically the ones where the gradient is non-vanishing. Within the setting of this work, it is thus natural to associate to I the discrete varifold

$$\mu_I = \sum_{\nabla I(x_i) \neq 0} \delta_{\left(x_i, \frac{\nabla I}{|\nabla I|}(x_i)\right)} \in \mathcal{D}.$$

It is straightforward that μ_I is invariant to increasing contrast changes. It also becomes invariant to decreasing ones by quotienting out the orientation of the unit gradient vectors, which in our framework is simply done by selecting an orientation-invariant kernel $\gamma(\langle d, d' \rangle)$ to define $\| \cdot \|_{W^*}$. In Figure 8, this approach is used to map two oppositely contrasted synthetic phantom brain images. We show both the alignment of the discrete varifolds as well as the full deformation applied to the image itself. Note that these images have no noise and a simple structure with relatively low number of non-vanishing gradients. There will be clearly the need for more validation to be done in the future in order to evaluate the practicality and robustness of this method for real multi-modal medical images.

6. Conclusion and future work

We have proposed, in this paper, a framework for large deformation inexact registration between discrete varifolds. It relies on the LDDMM setting for diffeomorphisms and include different models of group action on the space of varifolds. In each case, we derived the corresponding optimal control problems and the associated geodesic equations in Hamiltonian form. By combining those with the use of kernel-based fidelity metrics on varifolds, we proposed a geodesic shooting algorithm to numerically tackle the optimization problems. We finally illustrated the versatility and properties of this approach through examples of various natures which go beyond the classical cases of curves or surfaces.

Several improvements or extensions of this work could be considered for future work. From a theoretical standpoint, it would be for instance important to derive a more general 'continuous' varifold matching model i.e. with more general distributions than Dirac sums. Besides, higher dimensional varifolds could be

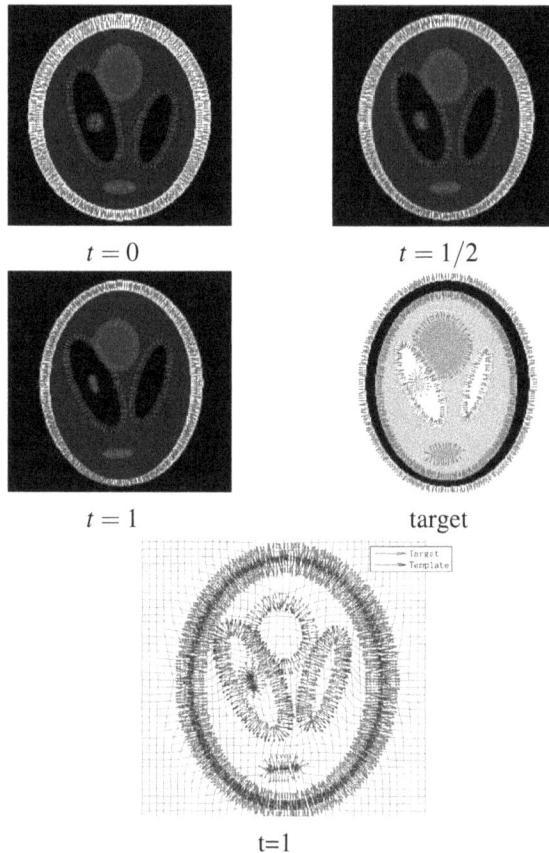

Fig. 8.   Registration of images modulo contrast changes. The matching is computed between the discrete varifolds associated to both images with the normalized action model and unoriented Gaussian fidelity term. The estimated deformation can be then applied to the template image.

possibly introduced within our model, although this would involve dealing with direction elements in Grassmann manifolds as in [7] instead of the simpler $\mathbb{S}^{n-1}$. Lastly, future work will also include adapting the existing fast GPU implementations for LDDMM to the new dynamical systems appearing here, with the objective of making the whole approach more scalable to real data applications.

Acknowledgments

The authors would like to thank Prof. Sarang Joshi for many enriching discussions that initiated parts of this work. This material is based upon work supported by the National Science Foundation under Grant No 1819131.

Appendix

Proof of Proposition 3.2

Let $\left(X_i(t), u_i(t), P_i^{(1)}(t), P_i^{(2)}(t)\right)$ and $V_t(\cdot)$ satisfy equations (3.8) and (3.9), then it is straightforward to verify that

$$\left(x_i(t), d_i(t), r_i(t), p_i^{(1)}(t), p_i^{(2)}(t), p_i^{(3)}(t)\right)$$
$$:= \left(X_i(t), \overline{u}_i(t), |u_i(t)|, P_i^{(1)}(t), |u_i(t)|P_i^{(2)}(t), \left\langle P_i^{(2)}(t), \overline{u}_i(t)\right\rangle\right)$$

is a solution of equation (3.5) for V with the initial conditions

$$(X_i(0), \overline{u}_i(0), |u_i(0)|, P_i^{(1)}(0), |u_i(0)|P_i^{(2)}(0), \langle P_i^{(2)}(0), \overline{u}_i(0)\rangle).$$

Moreover, we see that

$$\left\langle p_i^{(2)}(t), d_i(t)\right\rangle = \left\langle u_i(t), P_i^{(2)}(t)\right\rangle = r_i(t)p_i^{(3)}(t), \ \forall t \qquad (A.1)$$

which leads to V_t being equal to the vector field v_t defined in (3.6) and therefore to a solution for the system (3.5).

Conversely, let $\left(x_i(t), d_i(t), r_i(t), p_i^{(1)}(t), p_i^{(2)}(t), p_i^{(3)}(t)\right)$ and $v_t(\cdot)$ satisfying (3.5) and (3.6) with initial conditions such that

$$\left\langle p_i^{(2)}(0), d_i(0)\right\rangle = r_i(0)p_i^{(3)}(0). \qquad (A.2)$$

Now let $\left(X_i(t), u_i(t), P_i^{(1)}(t), P_i^{(2)}(t)\right)$ be the solution of (3.8) with the initial condition

$$\left(x_i(0), r_i(0)d_i(0), p_i^{(1)}(0), p_i^{(2)}(0)\right)$$

and vector field $v_t(\cdot)$. We define $V_t(\cdot)$ as in (3.9), then as in previous discussion, we see that

$$\left(X_i(t), \overline{u}_i(t), |u_i(t)|, P_i^{(1)}(t), |u_i(t)|P_i^{(2)}(t), \left\langle P_i^{(2)}(t), \overline{u}_i(t)\right\rangle\right)$$

is the solution for (3.5) with initial value

$$\left(X_i(0), \overline{u}_i(0), |u_i(0)|, P_i^{(1)}(0), |u_i(0)|P_i^{(2)}(0), \left\langle P_i^{(2)}(0), \overline{u}_i(0)\right\rangle\right)$$
$$= \left(x_i(0), d_i(0), r_i(0), p_i^{(1)}(0), p_i^{(2)}(0), p_i^{(3)}(0)\right).$$

Since $\left(x_i(t), d_i(t), r_i(t), p_i^{(1)}(t), p_i^{(2)}(t), p_i^{(3)}(t)\right)$ is a solution for the same initial value problem, by uniqueness of ODE, we obtain

$$\left(x_i(t), d_i(t), r_i(t), p_i^{(1)}(t), p_i^{(2)}(t), p_i^{(3)}(t)\right)$$
$$= \left(X_i(t), \overline{u}_i(t), |u_i(t)|, P_i^{(1)}(t), |u_i(t)|P_i^{(2)}(t), \left\langle P_i^{(2)}(t), \overline{u}_i(t)\right\rangle\right), \ \forall t \in [0,1].$$

Also, we have equation (A.1), and from this equation we have

$$P_{d_k^\perp}(p_k^{(2)}) + p_k^{(3)} r_k d_k = p_k^{(2)} + \left(\left\langle p_k^{(2)}, d_k\right\rangle - p_k^{(3)} r_k\right) d_k = p_k^{(2)}$$

and hence

$$v_t(\cdot) = \sum_{k=1}^P K(x_k, \cdot) p_k^{(1)} + D_1 K(x_k, \cdot)(d_k, p_k^{(2)})$$

$$= \sum_{k=1}^P K(X_k, \cdot) P_k^{(1)} + D_1 K(X_k, \cdot)(u_k, P_k^{(2)}) = V_t(\cdot).$$

Reduced Hamiltonian equations

For convenience, let us denote $f(|x_k - x_i|^2)$ by f_{ki} for any function f. Then the reduced Hamiltonian equations for the normalized action can be shown to be

$$
\begin{cases}
\dot{x}_i = \sum_{k=1}^P h_{ki} p_k^{(1)} + 2\dot{h}_{ki}\langle x_k - x_i, d_k\rangle P_{d_k^\perp}(p_k^{(2)}) \\[2mm]
\dot{d}_i = \sum_{k=1}^P -2\dot{h}_{ki}\langle x_k - x_i, d_i\rangle P_{d_k^\perp}\left(p_k^{(1)}\right) \\[2mm]
\qquad - \left[4\ddot{h}_{ki}\langle x_k - x_i, d_k\rangle\langle x_k - x_i, d_i\rangle + 2\dot{h}_{ki}\langle d_k, d_i\rangle\right] P_{d_k^\perp}(p_k^{(2)}) \\[2mm]
\dot{p}_i^{(1)} = \sum_{k=1}^P \Bigg\{ \left[2\dot{h}_{ki}\langle p_k^{(1)}, p_i^{(1)}\rangle + 4\ddot{h}_{ki}\langle x_k - x_i, d_k\rangle\langle P_{d_k^\perp}(p_k^{(2)}), p_i^{(1)}\rangle\right] \\[2mm]
\qquad - \Bigg[4\ddot{h}_{ki}\langle x_k - x_i, d_i\rangle\langle p_k^{(1)}, P_{d_i^\perp}(p_i^{(2)})\rangle \\[2mm]
\qquad + \left(8h_{ki}^{(3)}\langle x_k - x_i, d_k\rangle\langle x_k - x_i, d_i\rangle + 4\ddot{h}_{ki}\langle d_k, d_i\rangle\right)\langle P_{d_k^\perp}(p_k^{(2)}), P_{d_i^\perp}(p_i^{(2)})\rangle\Bigg]\Bigg\} \\[2mm]
\qquad \times (x_k - x_i) \\[2mm]
\qquad + \left[2\dot{h}_{ki}\langle P_{d_k^\perp}(p_k^{(2)}), p_i^{(1)}\rangle - 4\ddot{h}_{ki}\langle x_k - x_i, d_i\rangle\langle P_{d_k^\perp}(p_k^{(2)}), P_{d_i^\perp}(p_i^{(2)})\rangle\right] d_k \\[2mm]
\qquad - \left[2\dot{h}_{ki}\langle p_k^{(1)}, P_{d_i^\perp}(p_i^{(2)})\rangle + 4\ddot{h}_{ki}\langle x_k - x_i, d_k\rangle\langle P_{d_k^\perp}(p_k^{(2)}), P_{d_i^\perp}(p_i^{(2)})\rangle\right] d_i \\[2mm]
\dot{p}_i^{(2)} = \sum_{i=1}^P \left[2\dot{h}_{ki}\langle p_k^{(1)}, P_{d_i^\perp}(p_i^{(2)})\rangle + 4\ddot{h}_{ki}\langle x_k - x_i, d_k\rangle\left\langle P_{d_k^\perp}(p_k^{(2)}), P_{d_i^\perp}(p_i^{(2)})\right\rangle\right] \\[2mm]
\qquad \times (x_k - x_i) \\[2mm]
\qquad + 2\dot{h}_{ki}\left\langle P_{d_k^\perp}(p_k^{(2)}), P_{d_i^\perp}(p_i^{(2)})\right\rangle d_k \\[2mm]
\qquad - \left\langle d_i, p_i^{(2)}\right\rangle \left\{2\dot{h}_{ki}\langle x_k - x_i, d_i\rangle p_k^{(1)} + \left[4\ddot{h}_{ki}\langle x_k - x_i, d_k\rangle\langle x_k - x_i, d_i\rangle\right.\right. \\[2mm]
\qquad \left.\left. + 2\dot{h}_{ki}\langle d_k, d_i\rangle\right] P_{d_k^\perp}(p_k^{(2)})\right\} \\[2mm]
\qquad - \left\{2\dot{h}_{ki}\langle x_k - x_i, d_i\rangle\langle p_k^{(1)}, d_i\rangle\right. \\[2mm]
\qquad \left. + \left[4\ddot{h}_{ki}\langle x_k - x_i, d_k\rangle\langle x_k - x_i, d_i\rangle + 2\dot{h}_{ki}\langle d_k, d_i\rangle\right]\langle P_{d_k^\perp}(p_k^{(2)}), d_i\rangle\right\} p_i^{(2)}
\end{cases}
$$

In the pushforward action case, these equations are:

$$
\begin{cases}
\dot{x}_i &= \sum_{k=1}^{P} h_{ki} p_k^{(1)} + 2\dot{h}_{ki}\langle x_k - x_i, u_k \rangle p_k^{(2)} \\
\dot{u}_i &= \sum_{k=1}^{P} -2\dot{h}_{ki}\langle x_k - x_i, u_i \rangle p_k^{(1)} - \left[4\ddot{h}_{ki}\langle x_k - x_i, u_i \rangle\langle x_k - x_i, u_k \rangle + 2\dot{h}_{ki}\langle u_i, u_k \rangle \right] \\
& \quad \times p_k^{(2)} \\
\dot{p}_i^{(1)} &= \sum_{k=1}^{P} \Bigg\{ \left[2\dot{h}_{ki}\left\langle p_k^{(1)}, p_i^{(1)} \right\rangle + 4\ddot{h}_{ki}\langle x_k - x_i, u_k \rangle \left\langle p_k^{(2)}, p_i^{(1)} \right\rangle \right] \\
& \quad - \left[4\ddot{h}_{ki}\langle x_k - x_i, u_i \rangle\langle p_k^{(1)}, p_i^{(2)} \rangle \right. \\
& \quad + \left. \left(8h_{ki}^{(3)}\langle x_k - x_i, u_k \rangle\langle x_k - x_i, u_i \rangle + 4\ddot{h}_{ki}\langle d_k, d_i \rangle \right) \langle p_k^{(2)}, p_i^{(2)} \rangle \right] \Bigg\} (x_k - x_i) \\
& \quad + \left[2\dot{h}_{ki}\langle p_k^{(2)}, p_i^{(1)} \rangle - 4\ddot{h}_{ki}\langle x_k - x_i, u_i \rangle\langle p_k^{(2)}, p_i^{(2)} \rangle \right] u_k \\
& \quad - \left[2\dot{h}_{ki}\langle p_k^{(1)}, p_i^{(2)} \rangle + 4\ddot{h}_{ki}\langle x_k - x_i, u_k \rangle\langle p_k^{(2)}, p_i^{(2)} \rangle \right] u_i \\
\dot{p}_i^{(2)} &= \sum_{k=1}^{P} \left[2\dot{h}_{ki}\left\langle p_k^{(1)}, p_i^{(2)} \right\rangle + 4\ddot{h}_{ki}\langle x_k - x_i, u_k \rangle \left\langle p_k^{(2)}, p_i^{(2)} \right\rangle \right] (x_k - x_i) \\
& \quad + 2\dot{h}_{ki}\left\langle p_k^{(2)}, p_i^{(2)} \right\rangle u_k.
\end{cases}
$$

References

1. S. Arguillere, *The general setting of shape analysis*, preprint (2015).
2. S. Arguillere, E. Trélat, A. Trouvé, and L. Younes, *Shape deformation analysis from the optimal control viewpoint*, Journal de Mathématiques Pures et Appliquées **104** (2015), no. 1, 139–178.
3. B. Avants, C. Epstein, M. Grossman, and J. Gee, *Symmetric diffeomorphic image registration with cross-correlation: Evaluating automated labeling of elderly and neurodegenerative brain*, Medical Image Analysis **12** (2008), no. 1, 26–41.
4. M. F. Beg, M. I. Miller, A. Trouvé, and L. Younes, *Computing large deformation metric mappings via geodesic flows of diffeomorphisms*, International Journal of Computer Vision **61** (2005), no. 139–157.
5. Y. Cao, M. Miller, R. Winslow, and L. Younes, *Large deformation diffeomorphic metric mapping of vector fields*, IEEE Transactions on Medical Imaging **24** (2005), no. 9, 1216–1230.
6. C. Carmeli, E. De Vito, A. Toigo, and V. Umanita, *Vector valued reproducing kernel Hilbert spaces and universality*, Analysis and Applications **8** (2010), no. 01, 19–61.
7. N. Charon and A. Trouvé, *The varifold representation of non-oriented shapes for diffeomorphic registration*, SIAM Journal of Imaging Sciences **6** (2013), no. 4, 2547–2580.
8. J. Du, A. Goh, and A. Qiu, *Diffeomorphic Metric Mapping of High Angular Resolution Diffusion Imaging Based on Riemannian Structure of Orientation Distribution Functions*, IEEE Transactions on Medical Imaging **31** (2012), no. 5, 1021–1033.
9. S. Durrleman, P. Fillard, X. Pennec, Alain Trouvé, and Nicholas Ayache, *Registration, atlas estimation and variability analysis of white matter fiber bundles modeled as currents*, NeuroImage **55** (2010), no. 3, 1073–1090.
10. J. Glaunès, A. Qiu, M.I. Miller, and L. Younes, *Large deformation diffeomorphic metric curve mapping*, Int J Comput Vis **80** (2008), no. 3, 317–336.

11. J. Glaunès, A. Trouvé, and L. Younes, *Diffeomorphic matching of distributions: A new approach for unlabelled point-sets and sub-manifolds matching*, CVPR **2** (2004), 712–718.
12. J. Glaunès and M. Vaillant, *Surface matching via currents*, Proceedings of Information Processing in Medical Imaging (IPMI), Lecture Notes in Computer Science **3565** (2006), no. 381–392.
13. U. Grenander, *General pattern theory: A mathematical study of regular structures*, Clarendon Press Oxford, 1993.
14. H. Jacobs and S. Sommer, *Higher-order spatial accuracy in diffeomorphic image registration*, Geometry, Imaging and Computing **1** (2014), no. 4, 447–484.
15. S. Joshi and M. Miller, *Landmark matching via large deformation diffeomorphisms*, IEEE Transactions on Image Processing **9** (2000), no. 8, 1357–1370.
16. I. Kaltenmark, B. Charlier, and N. Charon, *A general framework for curve and surface comparison and registration with oriented varifolds*, Computer Vision and Pattern Recognition (CVPR) (2017).
17. L. Kühnel and S. Sommer, *Computational Anatomy in Theano*, Graphs in Biomedical Image Analysis, Computational Anatomy and Imaging Genetics: proceedings of the MFCA Workshop (2017), 164–176.
18. C. Ragni, N. Diguet, J-F. Le-Garrec, et al., *Amotl1 mediates sequestration of the Hippo effector Yap1 downstream of Fat4 to restrict heart growth*, Nature Communications **8** (2017).
19. D. Rueckert, L. I. Sonoda, C. Hayes, D. L. G. Hill, M. O. Leach, and D. J. Hawkes, *Nonrigid registration using free-form deformations: application to breast MR images*, IEEE Transactions on Medical Imaging **18** (1999), no. 8, 712–721.
20. S. Sommer, M. Nielsen, S. Darkner, and X. Pennec, *Higher-Order Momentum Distributions and Locally Affine LDDMM Registration*, SIAM Journal on Imaging Sciences **6** (2013), no. 1, 341–367.
21. F-X. Vialard, L. Risser, D. Rueckert, and C.J. Cotter, *Diffeomorphic 3D Image Registration via Geodesic Shooting Using an Efficient Adjoint Calculation*, International Journal of Computer Vision **97** (2012), no. 2, 229–241.
22. L. Younes, *Shapes and diffeomorphisms*, Springer, 2010.

Stochastic Metamorphosis with Template Uncertainties

Alexis Arnaudon[*] and Darryl D. Holm[†]

Department of Mathematics, Imperial College, London SW7 2AZ, UK
[*] alexis.arnaudon@imperial.ac.uk
[†] d.holm@imperial.ac.uk

Stefan Sommer

Department of Computer Science (DIKU), University of Copenhagen,
DK-2100 Copenhagen E, Denmark
sommer@di.ku.dk

In this paper, we investigate two stochastic perturbations of the metamorphosis equations of image analysis, in the geometrical context of the Euler-Poincaré theory. In the metamorphosis of images, the Lie group of diffeomorphisms deforms a template image that is undergoing its own internal dynamics as it deforms. This type of deformation allows more freedom for image matching and has analogies with complex fluids when the template properties are regarded as order parameters. The first stochastic perturbation we consider corresponds to uncertainty due to random errors in the reconstruction of the deformation map from its vector field. We also consider a second stochastic perturbation, which compounds the uncertainty of the deformation map with the uncertainty in the reconstruction of the template position from its velocity field. We apply this general geometric theory to several classical examples, including landmarks, images, and closed curves, and we discuss its use for functional data analysis.

Contents

1. Introduction

Variability in shapes can be modelled using flows of the group G of diffeomorphic deformations of the ambient domain Ω in which the shape is embedded. This is the basis of the large deformation diffeomorphic metric mapping (LDDMM) framework, see [30, 8, 9, 4]. In the LDDMM approach, the shape of an embedded template image $\eta \in N$ in the manifold of embedded shapes $\mathrm{Emb}(N, \Omega)$ changes via the action $g_t.\eta$ of time-dependent diffeomorphisms $g_t \in G$ on $\eta \in N$, through the action of g_t on the domain Ω. The metamorphosis extension [15, 22, 32, 33] of LDDMM introduces a further time-dependent variation η_t of the template to model the combined dynamics $g_t.\eta_t$.

   In this paper, we combine the geometrical metamorphosis framework of [15] with recent developments in stochastically perturbed Euler-Poincaré dynamics in fluid dynamics and shape analysis [13, 3, 2], to model evolutions of both shape and template under stochastic perturbations. The resulting framework allows modelling of random evolutions of shape and template simultaneously. A potential application of such an evolution is in modelling the progression of disease using computational anatomy, in which the model would address the analysis of disease progression in both the population average and in the individual. From longitudinal image data, mean evolutions over the population can be inferred. While average template evolutions can be modelled deterministically, models for the dynamics of each individual subject that include stochastic uncertainty are arguably more realistic than models supporting only deterministic trajectories. The stochastic metamorphosis model includes such non-deterministic variations by incorporating stochastic perturbations in shape and template simultaneously.

   One advantage of having a probabilistic approach for these problems is to allow for robust statistical analysis of shape deformations, which is not possible with an exact deterministic approach. Combined modelling of template and shape variations also appear in functional data analysis, where phase and amplitude variations interact. This is exemplified in the modelling of patterns of population growths for children where variation in height (here amplitude or template variation) is present and interact with variation in age (here phases) of the child growth. Deformation of the growth time axis represents variation in absolute age at which the phases of the growth process occur.

In metamorphosis dynamics, an analogy with the flows of complex fluids arises. In complex fluids, a diffeomorphic flow carries an order parameter, defined as a coset space for a broken symmetry of homogeneous fluids, on which the diffeomorphisms act. The order parameter moves with the fluid, but it can also have its own internal dynamics, which in turn is coupled to the fluid motion [12, 11]. A similar combined dynamics of shape and template also appears in the framework of geometrical shapes carrying functional information (Fshapes) [6].

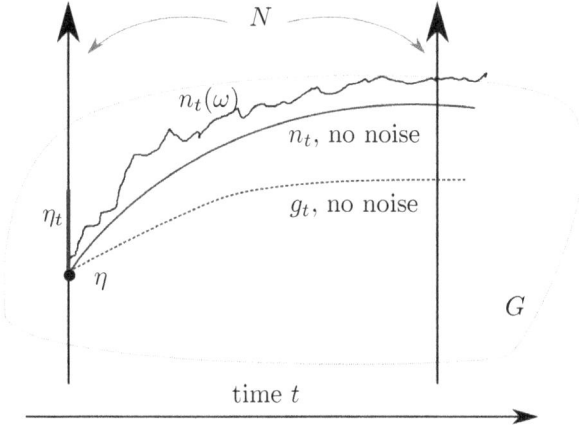

Fig. 1. (Color online) Sketch of the deformation group G, the shape space N (vertical arrows), evolution of the deformation variable g_t, template variable η_t, and shape variable n_t without noise ($W_t = 0$, blue), and shape variable n_t with noise ω (black). The shape space is illustrated as being linear (e.g. landmarks, images). However, the framework applies to general nonlinear shape spaces (e.g. curves, tensor fields).

1.1. *Background*

The LDDMM framework models the change of a shape $\eta \in N$ by the action of time-dependent flows of diffeomorphisms $g_t \in G$ on the embedding space Ω. One lifts the shape trajectory to a time-dependent curve g_t on the diffeomorphisms by setting $n_t = g_t.\eta \in N$. For a right invariant metric on the tangent space of a subgroup G of the diffeomorphism group $\mathrm{Diff}(N)$, N being the shape space, an energy can be defined as $E(g_t) = \int_0^1 \|\partial_t g_t\|_{g_t}^2 \mathrm{d}t$. Combined with a data attachment term, this approach allows matching of shapes and image registration [4]. The invariance of $E(g_t)$ under the right action of G implies that the metric descends to a metric structure on the data space N itself. The action of g_t is different for each data type, but otherwise, the mathematical framework is formally equivalent for different classes of shapes. The use of the flows g_t to model the shape variability is fundamental and the right trivialization $v_t := \partial_t g_t \circ g_t^{-1}$ gives a Eulerian interpretation of the metric. The right invariance of the metric enables Euler-Poincaré reduction of the dynamics to the dual of the Lie algebra of G to be performed, and the critical paths for E appear from the reduced dynamics.

Metamorphosis extends the LDDMM setting by letting the template vary in time as well as the deformation, thereby resulting in the flow $g_t.\eta_t$, in which $\eta_0 = \eta$ is the original template. The metamorphosis energy is encoded into a Lagrangian depending on both the G and N variability, again assuming invariance of the energy to the group action on both G and N. A particular example of metamorphosis dynamics arises in image analysis, where the image I_t changes both by deformation via the right action $g_t.I_t = I_t \circ g_t^{-1}$ and via a pointwise change $\partial_t I_t(x)$ for each pixel/voxel x.

In [3, 2], a stochastic model of shape evolution was introduced that preserves the Euler-Poincaré theory of the deterministic LDDMM framework. The model is based on the stochastic fluid dynamics model [13] where right-invariant noise is introduced to perturb the reconstruction equation that evolves the flow from the reduced dynamics. In deterministic LDDMM, the reconstruction equation specifies the evolution of the group element by $\partial_t g_t = v_t \circ g_t$ generated by the reduced Eulerian velocity vector field v_t. Stochasticity is introduced as a perturbation to the reconstruction equation, by introducing the stochastic time differential

$$\mathrm{d}g_t\, g_t^{-1} = v_t \mathrm{d}t + \sum_{l=1}^{N} \sigma_l \circ \mathrm{d}W_t^l. \tag{1.1}$$

Here W_t^l are standard Wiener processes and σ_l are vector fields on the data domain which characterize the spatial correlation of the noise. As it turns out, the noise in Stratonovich form is denoted conventionally with the same symbol ($\circ$) that denotes composition of maps. This coincidence should not cause any confusion. However, just to be sure, we will write the composition of maps as concatenation whenever the two meanings appear in the same equation, as in (1.1). The stochastic perturbation is applied directly on the velocity field, or reduced variable in the context of reduction by symmetries, which implies that the resulting stochastic equation remains right-invariant, thus compatible with the original deterministic right-invariant LDDMM metric. This approach preserves many of the geometric structures of the deterministic framework. Importantly, the descent of the stochastic model to particular data types is similar to the way the metric descends in the deterministic LDDMM framework.

A stochastic metamorphosis extension of the stochastic Euler-Poincaré framework was introduced in [14]. The stochastic perturbations there were also introduced in the reduced variable influencing the deformation flow from the reconstruction equation. The template evolution η_t is still deterministic. The aim of the present paper is to extend this model to include noise in the template evolution η_t as well. We will make this extension on the reduced template velocity $g_t \partial_t \eta_t$ similarly to the perturbation of the group variable. This procedure results in simultaneous stochastic perturbations of the flow equations for both g_t and η_t.

1.2. *Paper outline*

After a brief survey of the deterministic metamorphosis framework in section 2.1, we formally derive the stochastic model in section 2.2. We then show in section 2.3 how to derive these equations in the Hamilton-Pontryagin formulation, where the noise appears as a stochastic constraint in the variational principle. We end the theoretical section by deriving the corresponding Hamiltonian stochastic equations in (2.4) to then move to some classical examples of image analysis and computational anatomy in section 3, including landmarks and images. The inclusion of two types of stochastic variations links the framework to combined random phase and amplitude variations in functional data analysis. We provide perspectives of the method to future applications in functional data analysis and computational anatomy in section 4.

2. General stochastic metamorphosis

In this section, we introduce the stochastic deformation of metamorphosis, but first, we recall the basis of this theory, in the context of reduction by symmetry. We will only review what will be needed for our exposition, and we refer to [15] and [14] for more extensive treatments.

2.1. *Deterministic metamorphosis*

The theory of metamorphosis begins with a template N, considered here as a manifold of shape representations, such as landmarks or images, upon which a group of diffeomorphism $G = \text{Diff}(N)$ acts. The parameter space of this theory is $G \times N$, with curves $(g_t, \eta_t) \in G \times N$, where g_t is the deformation curve and η_t is the template curve. The image curve will be denoted $n_t = g_t.\eta_t \in N$, where the dot represents the group action. This curve is the total motion of the template, or image N, under both the deformation and its own dynamics. For standard LDDMM, the motion of the image is only $n_t = g_t.\eta$, for a fixed reference template η. This combined action thus allows more freedom in the matching procedure, while remaining compatible with the theory of reduction by symmetries, which we now describe. We first define the two reduced velocity fields as

$$u_t := \dot{g}_t g_t^{-1}, \quad \text{and} \quad \nu_t := g_t \dot{\eta}_t. \tag{2.1}$$

The first is the reduced deformation velocity and the second is the reduced template velocity. We then assume that the original Lagrangian of this theory is invariant under the group action of G, so that we may write the reduced Lagrangian in terms of the reduced velocity fields and the image position n_t, i.e.,

$$L(g_t, \dot{g}_t, \eta_t, \dot{\eta}_t) = l(u_t, n_t, \nu_t). \tag{2.2}$$

Because the reduced Lagrangian still depends on the template variable n_t, reduction by the action of the diffeomorphisms will result in a semi-direct product structure, where the template is an advected quantity, in the language of fluid dynamics.

We next compute the variations of the three variables in the reduced Lagrangian, upon introducing the notation $\xi_t = \delta g_t g_t^{-1}$ and $\omega = g\delta\eta$, where δg and $\delta \eta$ are free variations, to obtain

$$\delta u = \dot{\xi}_t - [u_t, \xi_t],$$
$$\delta n = \omega_t + \xi_t n_t, \tag{2.3}$$
$$\delta \nu = \dot{\omega}_t + \xi_t \nu_t - u_t \omega_t.$$

In these formulas, we need to specify what we mean by the multiplication, as the vector fields live in different spaces. In fact, $u_t, \xi_t \in \mathfrak{g} = \mathfrak{X}(N)$ are vector fields; so the Lie bracket is the natural operation. Recall that $\eta_t \in N$, thus $\xi_t \eta_t$ corresponds to the tangent map of the action of G on the manifold N, and similarly for $\nu \in TN$, where the action is on the tangent space of N. We do not need these actions explicitly now, but we will need their 'adjoint action' in the following sense:

$$\langle n^* \diamond m, u \rangle_{\mathfrak{g}} = -\langle n^*, un \rangle_N \,, \tag{2.4}$$

$$\langle u \star \nu^*, \nu \rangle_{TN} = \langle \nu^*, u\nu \rangle_{TN} \,, \tag{2.5}$$

where $N \in N$, $n^* \in N^*$, $\nu \in TN$, $\nu^* \in T^*N$ and $u \in \mathfrak{g}$. The first equality defines the diamond operation ($\diamond$), which will serve as a force term to capture the coupling between the advected quantity n and the main dynamics of the diffeomorphism group. The second equality defines the star operation ($\star$), which is the adjoint of the action of u on TN. That is, it defines the action of u on T^*N.

Applying the variational calculus to the action $S = \int l(u_t, n_t, \nu_t) dt$, we obtain the Euler-Poincaré formulation of the metamorphosis equation in the form

$$\frac{d}{dt} \frac{\delta l}{\delta u} + \mathrm{ad}^*_{u_t} \frac{\delta l}{\delta u} + \frac{\delta l}{\delta n} \diamond n + \frac{\delta l}{\delta \nu} \diamond \nu = 0 \,,$$
$$\frac{d}{dt} \frac{\delta l}{\delta \nu} + u_t \star \frac{\delta l}{\delta \nu} - \frac{\delta l}{\delta n} = 0 \,, \tag{2.6}$$

together with the reconstruction equation

$$\dot{n} = u_t n_t + \nu_t \,. \tag{2.7}$$

We refer to [15, 14] for the details of this derivation, which we will do in the context of Hamilton-Pontryagin with noise in the next section.

From here, a choice of Lagrangian and data N will reduce the system to particular cases, some of which we discuss in the applications section 3.

2.2. *Formal derivation of the stochastic equations*

We will first derive the equation informally, using 'stochastic variations', then show a more straightforward derivation using the Hamilton-Pontryagin principle. The second derivation also has the advantage of revealing the effects of the noise more transparently.

In order to introduce a noise compatible with the Euler-Poincaré equation, we need to perturb the theory at its core, which is, in this case, the definition of the reduced velocities in (2.1). Indeed, the variations were computed from these definitions, and the deterministic Euler-Poincaré equation emerged. Although a single relation is used in the Euler-Poincaré equation (2.7), we will split it into two parts, and perturb them with two different noise components as follows,

$$dg_t g_t^{-1} = u_t(x)dt + \sum_{l=1}^{K^u} \sigma_l^u(x) \circ dW_t^l =: du_t(x),$$

$$g_t d\eta = \nu_t dt + \sum_{k=1}^{K^\nu} \sigma_k^\nu \circ dW_t^k =: d\nu_t.$$

(2.8)

In a slight abuse of notation, $du_t(x)$ and $d\nu_t$ are written as stochastic processes. Here $\sigma_l^u : N \to \mathfrak{g}$ are a set of K^u vector fields on the domain Ω, and $\sigma_l^\nu \in TN$ are another set of K^ν tangent vectors on the template. We also denote by W_t^l or W_t^k the $K^u + K^\nu$ independent standard Wiener processes, see for example [25] for more on this process. In addition, we denote by $x_0 \in \Omega$ the Lagrangian labels upon which g_t acts, so that the first equation can be written equivalently as

$$dg_t = u_t(g_t x_0)dt + \sum_{l=1}^{K^u} \sigma_l^u(g_t x_0) \circ dW_t^l.$$

The second equation for η in (2.8) does not have any x_0 dependence, as it is an equation for the template itself. Thus, σ_k^ν are not functions of N; rather, they are tangent vectors to N.

With the notation for du_t and $d\nu_t$ in (2.8), we have the complete reconstruction relation for the stochastic image template n_t

$$dn_t = du_t \, n_t + d\nu_t.$$

(2.9)

Because $n_t \in N$, the concatenation $du_t \, n_t$ means the composition $du_t(n_t)$. In (2.9), the noise in the u_t vector field was introduced in [14], based on the stochastic fluid dynamics model of [13], whereas the noise in the ν_t field is new. The first noise term in (2.9) corresponds to random errors in the reconstruction of the diffeomorphism path from its velocity field, while the second one represents random errors for the reconstruction of the template position from its velocity field. In stochastic metamorphosis, the two noise terms will affect the dynamical equations differently.

From these stochastic perturbations of the reconstruction relation, we can formally compute the variations and obtain

$$\delta u = \mathrm{d}\xi_t + [\xi, \mathrm{d}u_t]\,,$$
$$\delta \nu = \mathrm{d}\omega + \xi \mathrm{d}\nu_t - \mathrm{d}u_t \omega\,.$$

(2.10)

These are convenient expressions, but they introduce the variations as stochastic processes; so they should not be taken at face value without further analysis. We will see in the next section how to re-derive these equations without introducing stochastic variations, by using the Hamilton-Pontryagin principle. Because the results are identical for the two methods, we can proceed formally here by using these variations as we did in the deterministic variational principle to obtain the following stochastic reduced metamorphosis equations in Euler-Poincaré form,

$$\mathrm{d}\frac{\delta l}{\delta u} + \mathrm{ad}^*_{\mathrm{d}u_t}\frac{\delta l}{\delta u} + \frac{\delta l}{\delta n}\diamond n\mathrm{d}t + \frac{\delta l}{\delta \nu}\diamond \mathrm{d}\nu_t = 0\,,$$
$$\mathrm{d}\frac{\delta l}{\delta \nu} + \mathrm{d}u_t \star \frac{\delta l}{\delta \nu} - \frac{\delta l}{\delta n}\mathrm{d}t = 0\,,$$

(2.11)

as well as equation (2.9), all to be compared with the deterministic case in equations (2.6) and (2.7).

2.3. Derivation using the Hamilton-Pontryagin principle

We now show how to rederive the stochastic metamorphosis equations more transparently, without introducing stochastic variations (2.10). For this purpose, we will use the stochastic Hamilton-Pontryagin approach and closely follow the exposition of [14].

The deterministic Hamilton-Pontryagin principle is a variational principle with the following constrained action

$$S(u_t, n_t, \dot{n}_t, \nu_t, g_t, \dot{g}_t) = \int_0^1 l(u_t, n_t, \nu_t)\mathrm{d}t + \int_0^1 \langle M_t, (\dot{g}_t g_t^{-1} - u_t)\rangle\mathrm{d}t$$
$$+ \int_0^1 \langle \sigma_t, (\dot{n}_t - \nu_t - u_t n_t)\rangle\mathrm{d}t\,,$$

(2.12)

where $M_t \in \mathfrak{X}^*(N)$ and $\sigma_t \in T^*N$ are generalised Lagrange multipliers to enforce the constraint of the reconstruction relations. Taking free variations for all the variables yields the deterministic reconstruction relation (2.7) and the deterministic Euler-Poincaré equation (2.6). We refer to [14] for more details of the derivation. The crucial point here is to allow free variations, by introducing constraints into the variational principle, and not in the variations as in the standard Euler-Poincaré reduction theory. An alternative approach would be to use the Clebsch constrained variational method used for fluid dynamics in [13].

In the present context, we enforce the stochastic reconstruction relations (2.8) via the following stochastic Hamilton-Pontryagin principle

$$S(u_t, n_t, dn_t, \nu_t, g_t, dg_t) = \int_0^1 l(u_t, n_t, \nu_t) + \int_0^1 \langle M_t, (dg_t g_t^{-1} - du_t) \rangle$$
$$+ \int_0^1 \langle \sigma_t, (dn_t - d\nu_t - du_t n_t) \rangle, \quad (2.13)$$

or, more explicitly, upon substituting for du_t and $d\nu_t$ from (2.8), we have

$$S(u_t, n_t, dn_t, \nu_t, g_t, dg_t) = \int_0^1 l(u_t, n_t, \nu_t) dt$$
$$+ \int_0^1 \left\langle M_t, dg_t g_t^{-1} - u_t dt - \sum_{l=1}^{K^u} \sigma_l^u(x) \circ dW_t^l \right\rangle$$
$$+ \int_0^1 \left\langle \sigma_t, dn_t - \nu_t dt - \sum_{k=1}^{K^\nu} \sigma_k^\nu(x) \circ dW_t^k \right\rangle$$
$$- \int_0^1 \left\langle \sigma_t, \left(u_t dt + \sum_{l=1}^{K^u} \sigma_l^u(x) \circ dW_t^l \right) n_t \right\rangle. \quad (2.14)$$

Proposition 2.1: *The stochastic variational principle $\delta S = 0$ with action (2.14) yields the stochastic Euler-Poincaré equation (2.11) with stochastic reconstruction relation (2.8) and (2.9).*

Proof: The proof is a direct computation by taking free variations. We will show the key steps below. First, the variations with respect to M_t and σ_t yield the reconstruction relations (2.8) and (2.9). Then, the variations with respect to u_t, n_t and ν_t specify

$$\frac{\delta l}{\delta u_t} = M_t + \sigma_t \diamond n_t, \quad \text{and} \quad \frac{\delta l}{\delta \nu_t} = \sigma_t. \quad (2.15)$$

We also have, for the n_t variations,

$$\frac{\delta l}{\delta n_t} dt = d\sigma_t + u_t \star \sigma_t dt + \sum_{l=1}^{K^u} \sigma_l^u(x) \star \sigma_t \circ dW_t^l. \quad (2.16)$$

Finally, for $\xi = \delta g g^{-1}$ vanishing at the endpoints, we have

$$\delta(\mathrm{d} g_t g_t^{-1}) = \mathrm{d}\xi - \left[u_t \mathrm{d}t + \sum_{l=1}^{K^u} \sigma_l^u(x) \circ \mathrm{d}W_t^l, \xi \right]. \qquad (2.17)$$

From this computation, we have the last term in the calculus of variations which reads

$$\mathrm{d}M_t = -\operatorname{ad}_{u_t}^* M_t - \sum_{l=1}^{K^u} \operatorname{ad}_{\sigma_l^u(x)}^* M_t \circ \mathrm{d}W_t^l. \qquad (2.18)$$

Finally, substituting the values of M_t and σ_t of (2.15) in equation (2.16) and (2.18) yields the stochastic metamorphosis equation (2.11) after a few more manipulations (see Corollary 3 of [14]). $\square$

2.4. *Hamiltonian formulation*

Provided that the Lagrangian is hyper-regular, the stochastic metamorphosis equation (2.11) can be written as a stochastic Hamiltonian equation with Hamiltonian obtained via the reduced Legendre transform,

$$h(\mu, \sigma, n) = \langle \mu, u \rangle + \langle \sigma, \nu \rangle - l(u, \nu, n), \qquad (2.19)$$

in which μ and σ_t are the conjugate variables of u_t and ν_t, respectively. The noise is encoded into the stochastic potentials

$$\Phi_l^u(\mu_t) = \langle \mu_t, \sigma_l^u \rangle_{\mathfrak{g} \times \mathfrak{g}^*}, \quad \text{and} \quad \Phi_k^\nu(\sigma_t) = \langle \sigma_t, \sigma_k^\nu \rangle_{TN \times T^*N}, \qquad (2.20)$$

such that the stochastic equation of motion has a Hamiltonian drift term with h and stochastic terms obtained via the same Hamiltonian structure, but with stochastic potentials. Notice that the two potentials have a different pairing, one on the Lie algebra of the diffeomorphism group, and the other on the tangent space of the template manifold. The Hamiltonian structure is given in [14] and we will only display here the Hamiltonian equations

$$\mathrm{d}\mu_t + \operatorname{ad}_{\frac{\delta h}{\delta \mu}}^* u \mathrm{d}t + \sigma \diamond \frac{\delta h}{\delta \sigma} \mathrm{d}t + \frac{\delta h}{\delta n} \diamond n \mathrm{d}t$$

$$+ \sum_l \operatorname{ad}_{\frac{\delta \Phi_l^u}{\delta \mu}}^* u \circ \mathrm{d}W_t^l + \sum_l \sigma_t \diamond \frac{\delta \Phi_l^\nu}{\delta \sigma} \circ \mathrm{d}W_t^l = 0, \qquad (2.21)$$

$$\mathrm{d}\sigma_t + \frac{\delta h}{\delta \mu} \star \sigma_t \mathrm{d}t - \frac{\delta h}{\delta n} + \sum_l \frac{\delta \Phi_l^u}{\delta \mu} \star \sigma_t \circ \mathrm{d}W_t^l = 0.$$

In the examples in the next section, we will use this formulation to derive the stochastic equations of motion. Taking the Hamiltonian approach turns out to be more transparent than the Lagrangian description.

3. Applications

Following [15], we explicitly provide the stochastic metamorphosis equations for a few classical examples, including landmarks and images, and leave other applications such as closed planar curves, densities or tensor fields for later works.

3.1. *Landmarks and peakons*

Consider the case when the template manifold N is the space of n landmarks $\mathbf{q} = (q_1, \ldots, q_n) \in \Omega^n$ with momenta $\mathbf{p} = (p_1, \ldots, p_n) \in T_{\mathbf{q}}\Omega^n \cong \Omega^n$. One needs to specify a Lagrangian for this system, and the simplest is

$$l(u, n, \nu) = \frac{1}{2}\|u\|_K^2 + \frac{\lambda^2}{2}\sum_{i=1}^n |p_i|^2\,, \tag{3.1}$$

where the first norm depends on the kernel $K(x)$ and the second norm is the vector norm of the momenta multiplied by a constant λ^2. In this case, we interpret the momenta as the conjugate variables to the template deformation vector field ν in order to have an equation only in term of the position and momenta of the landmarks. The derivation of the landmark equation is rather standard. Hence, we will only show it on the Hamiltonian side. We refer, for example, to [15] for more details of the deterministic derivation, or to [3] and [16] for discussions of the stochastic landmark dynamics.

Recall that the landmark Hamiltonian is

$$h_K(\mathbf{p}, \mathbf{q}) = \frac{1}{2}\sum_{ij} p_i \cdot p_j K(q_i - q_j)\,, \tag{3.2}$$

and the metamorphosis Hamiltonian is thus

$$h(q_i, p_i) = h_K(\mathbf{q}, \mathbf{p}) + \frac{\lambda^2}{2}\sum_{i=1}^n |p_i|^2\,. \tag{3.3}$$

The stochastic potentials (2.20) become in this case

$$\Phi_l^u(\mathbf{q}, \mathbf{p}) = \sum_i p_i \cdot \sigma_l^u(q_i) \qquad \text{and} \qquad \Phi_i^\nu(\mathbf{p}) = p_i \cdot \sigma_i^\nu\,. \tag{3.4}$$

Notice that the stochastic potential Φ^ν is described by a fixed vector, where σ_i^ν is the amplitude of the noise for the landmark i. However, for the stochastic potential Φ^u, we have to specify space (or $\mathbf{q}$) dependent functions $\sigma_l^u(\mathbf{q})$. This simple form comes from the fact that we used a discrete set of points and $\nu = \mathbf{p}$ for the template deformation, and the summation over k becomes a summation over the landmark index. In addition, a sum of two Wiener processes is another Wiener process with the sum of the amplitude (if it is additive and in Itô form). From this observation, one can see that the general equation $\Phi_k^\nu(\mathbf{p}) = \sum_i p_i \cdot \sigma_k^\nu$ is equivalent to a change of amplitudes σ_k^ν and $i = k$.

We compute the stochastic Hamiltonian equations for landmarks to arrive at

$$
\begin{aligned}
dq_i &= \frac{\partial h_K}{\partial p_i} dt + \sum_l \sigma_i^u \circ dW_t^l + \lambda^2 p_i dt + \sigma_i^\nu dW_t^i, \\
dp_i &= -\frac{\partial h_K}{\partial q_i} dt + \sum_l \partial_{q_i}(p_i \cdot \sigma_l^u) \circ dW_t^l,
\end{aligned}
\tag{3.5}
$$

in which we can use the Itô integral for the ν-noise, as it is additive.

Notice that setting $\lambda = 0$ recovers the standard landmark dynamics, but with an additive noise in the position equation. This is different from the conventional physical perspective, in which additive noise often appears in the momentum equation, as in [31, 35, 21].

3.2. *Images*

The present stochastic metamorphosis framework can be directly applied to images, by taking the template space N to be the space of smooth functions from the domain $\Omega \subset \mathbb{R}^2$ to $\mathbb{R}$. We set $u_t \in \mathfrak{X}(\Omega)$ the deformation vector field and $\rho \in TN \cong N$ the template vector field. As before, the Lagrangian must have two parts, and the simplest non-trivial one is the sum of kinetic energies written as

$$
l(u, n, \nu) = \frac{1}{2}\|u_t\|_K^2 + \frac{\lambda^2}{2}|\rho_t|_{L^2}^2,
\tag{3.6}
$$

where the first norm depends on the kernel K and the second norm is the standard L^2 norm over Ω. By choosing a L^2 norm we can identify ρ_t with its dual in the case $\lambda = 1$. We will thus not distinguish between σ_t and ν_t of the general framework.

Thus, as before, we use the Hamiltonian formulation of the stochastic metamorphosis equations with the stochastic potentials,

$$
\Phi_l^u(m_t) = \int_\Omega \langle m_t(x), \sigma_l^u(x) \rangle dx \quad \text{and} \quad \Phi_k^\nu(\sigma_t) = \int_\Omega \langle \rho_t(x), \sigma_l^\nu(x) \rangle dx.
\tag{3.7}
$$

Notice that in this case, both σ_l^u and σ_l^ν are functions of the domain Ω, and they encode spatial correlation structure of the stochastic perturbations.

Then, because the Hamiltonian structure has three sorts of terms, the ad*, the $\diamond$ and the $\star$ terms defined in equation (2.5), which in this case are

$$\mathrm{ad}^*_{u_t} m_t = (u_t \cdot \nabla)m_t + (m_t \cdot \nabla)u_t + \mathrm{div}(u_t)m_t \,,$$

$$\sigma_t \diamond \nu_t = \sigma_t \cdot \nabla \nu_t \,,$$

$$u_t \star \sigma_t = \nabla \cdot (\sigma_t u_t) \,,$$

we arrive at the following set of stochastic PDEs (for any λ)

$$\mathrm{d}m_t + \mathrm{ad}^*_{u_t} m_t \mathrm{d}t + \sum_l \mathrm{ad}^*_{\sigma^u_l} m_t \circ \mathrm{d}W^l_t = \lambda^2 \rho_t \cdot \nabla \rho_t \mathrm{d}t + \sum_k \rho_t \cdot \nabla \sigma^\nu_k \circ \mathrm{d}W^k_t \,,$$

$$\mathrm{d}\rho_t + \nabla \cdot (\rho_t u_t)\mathrm{d}t + \sum_l \nabla \cdot (\rho \sigma^u_l) \circ \mathrm{d}W^l_t = 0 \,.$$

$$(3.8)$$

Another important equation is the reconstruction relation (2.9), which now reads

$$\mathrm{d}g_t = u_t(g_t)\mathrm{d}t + \sum_l \sigma^u_l(g_t) \circ \mathrm{d}W^l_t + \rho_t \mathrm{d}t + \sum_k \sigma^\nu_k \circ \mathrm{d}W^k_t \,. \qquad (3.9)$$

Notice that if we set $\lambda = 1$, the effect of the density, or template motion on the momentum m only appears via the noise term, similarly to the landmark case.

In the one-dimensional case, the metamorphosis equation is known to reduce to the so-called CH2 system, which is equation coupling the Camassa-Holm equation with a density advection equation for $\rho_t = \nu_t$. We refer to [15, 7] and references therein for more details about this equation and its complete integrability in the deterministic case. A similar reduction holds for both stochastic deformations, and we have the following stochastic CH2 equation

$$\mathrm{d}m + (u\partial_x m + 2m\partial_x u)\mathrm{d}t$$

$$= -\rho\partial_x \rho \mathrm{d}t - \sum_k \rho\partial_x \sigma^\nu_k \circ \mathrm{d}W^k_t - \sum_l \left(\sigma^u_l \partial_x m + \sum_l 2m\partial_x \sigma^u_l \right) \circ \mathrm{d}W^l_t \,,$$

$$\mathrm{d}\rho + \partial_x(\rho u)\mathrm{d}t + \partial_x(\rho \sigma^u_l) \circ \mathrm{d}W^l_t = 0 \,.$$

$$(3.10)$$

Compared to the landmark example, the noise associated with the template dynamics is described by a set of functions of the image, not a set of fixed vectors. The difference between the nature of these two types of noise is thus less apparent, apart from how they appear in the equation.

4. Perspectives

4.1. *Computational anatomy*

Estimation of population atlases and longitudinal analysis of anatomical changes caused by disease progression constitute integral parts of computational anatomy [36]. The relation between these problems and the stochastic metamorphosis model presented here can be illustrated by the analysis of longitudinal brain MR-image data of patients suffering from Alzheimer's disease. The data manifold N is here a vector space of images as described above with $\Omega \subseteq \mathbb{R}^3$.

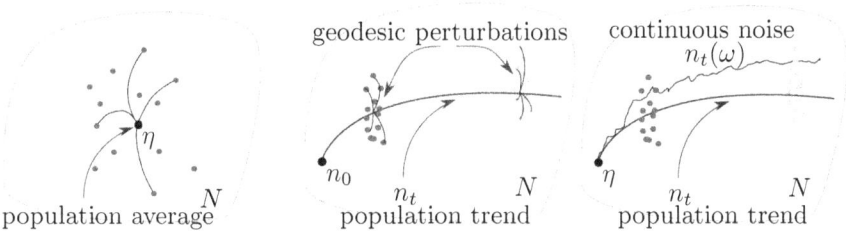

geodesic perturbations　　continuous noise $n_t(\omega)$

n_0　　N　　η　　N

N　　n_t　　n_t

population average　　population trend　　population trend

Fig. 2. (Color online) (left) Template estimation in the form (4.1) aims at finding a single descriptor η for the population average of the observed shapes $n^1, \ldots, n^k$ (red dots) in the nonlinear shape space N. The variational principle (4.1) corresponds to assuming n^i arise from geodesic perturbations of η. (center) Geodesic regression models a population trend as a geodesic n_t. Observations at different time points $(n_{t_1}^i$ red, $n_{t_2}^i$ green) arise as perturbations of the points n_{t_1} and n_{t_2} by random geodesics. (right) Stochastic metamorphosis models the evolution of the population trend n_t deterministically while observations $n_{t_j}^i = n_{t_j}(\omega^i)$ appear from individual noise realizations ω^i. The perturbations are time continuous and apply to each case i individually making the model natural for modelling longitudinal evolutions with noise.

Focusing first on template estimation, in medical imaging commonly denoted atlas estimation, the aim is to find a population average of data assumed observed at a fixed time point. In the literature, this is for example pursued by minimizing the total sum of the regularized LDDMM energies of deterministic geodesic trajectories that deform the atlas to match the observed data [17]. For k data points $n^1, \ldots, n^k$ and with data matching term $S : N \times N \to \mathbb{R}$, the template η is then estimated by joint minimization of

$$\min_{(\eta, v_t^1, \ldots, v_t^k)} \left(\sum_{i=1}^{k} \int_0^1 \|v_t^i\|^2 \mathrm{d}t + S(\phi_T^i.\eta, n^i) \right), \tag{4.1}$$

where the deformations ϕ_T^i each are endpoints of the integral of the vector fields v_t^i on an interval $[0, T]$.

A different approach to atlas estimation is to perform inference in statistical models where observations are assumed random perturbations of a template and inference of the template is performed via maximum-likelihood or maximum-a-posteriori estimation. This approach is pursued, for example, in a [1, 37, 26]. See also the discussion below.

Longitudinal analysis aims at capturing the average time evolution of the brain shape caused by the disease [23, 24]. A common approach here is to estimate a general deterministic trend that is perturbed by noise at discretely observed time points in order to describe the observed images [10]. For example, the noise can take the form of random initial velocity vectors for geodesics emanating from the deterministic trajectory.

The stochastic metamorphosis framework proposed here combines deterministic longitudinal evolution of the template in both shape, represented by the deformations g_t, and in the template image, $n_t = g_t.\eta_t$. We can assume longitudinal observations $n_{t_j}^i$, $i = 1, \ldots, k$, $j = 1, \ldots, t_l$ at l time points are realizations of the stochastic model with time-continuous noise process drawn for each subject i. The stochastic perturbations are thus tied to each subject affecting the dynamics simultaneously with the evolution of the deterministic flow. The relation between this model, geodesic regression models, and atlas estimation is illustrated in Figure 2.

Because of the randomness, algorithms for inference of the template η_t and its evolution $n_t = g_t.\eta_t$ from data can naturally be formulated by matching statistics of the data, e.g. by matching moments or by maximum-likelihood as done for the landmark case of stochastic EPDiff equations in [3]. Development of such inference schemes constitutes natural future research directions.

4.2. *Phase and amplitude in functional data analysis*

While images exhibit variations in both the intensity and shape of the image domain, signals in functional data analysis often exhibit combined variation in amplitude and phase. For a signal $f : I \to N$ defined on an interval I, amplitude variations refer to variations of the values $f(s)$ in N for each fixed $s \in I$ while phase variations cover changes in the parametrization of the domain I. This is illustrated with $N = \mathbb{R}$ in Figure 3. An example of such combined phase and amplitude variations is provided in the growth curves of children and young adults; in which phase variation is connected to the absolute height of the subject while phase variations arise from growth and growth spurts occurring at different ages for different children.

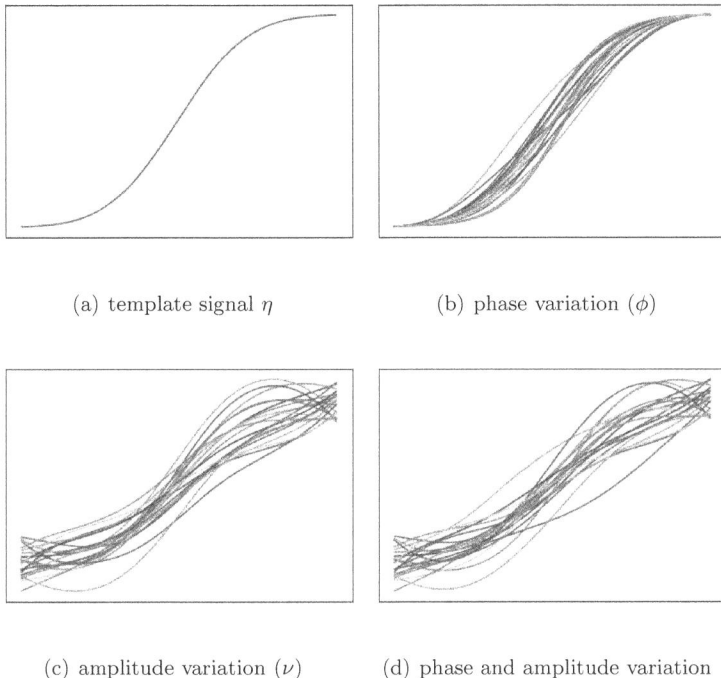

(a) template signal η

(b) phase variation (ϕ)

(c) amplitude variation (ν)

(d) phase and amplitude variation

Fig. 3. A template signal (a) can be perturbed by (b) variation in phase, in (4.2) denoted ϕ; (c) variation in amplitude, ν in (4.2); (d) phase and amplitude simultaneously.

Recent literature covers multiple approaches for identifying, separating and performing inference in situations with combined phase and amplitude variation [27, 20, 34]. One example of a generative model in this settings is the mixed-effects model [27, 19]

$$f(s) = \eta(\phi^{-1}(s)) + \nu(s) + \epsilon \, , \ s \in I \, , \tag{4.2}$$

where the average signal η is deformed in phase by the action $\phi.\eta = \eta \circ \phi^{-1}$ of a deformation ϕ of the interval I, and in amplitude by the additive term ν. Here η is considered a fixed, non-random effect while both ϕ and ν are random. Illustrated with the growth curve case above, η models the population average growth curve for each age s, while ϕ controls the timing of the growth process for the individual children and ν the absolute height difference to the population average. One observes that the model (4.2) is nonlinear, because of the coupling between ϕ and η. In addition, a model for the deformations ϕ is needed, and the randomness appearing in both ϕ and ν must be specified.

Whereas the LDDMM model is widely used in image analysis, this framework has not yet seen many applications for modelling deformations in functional data analysis, such as the phase variation appearing in (4.2). Instead, works such as [27] use a small-deformation model $\phi(s) = s + v(s)$ with random vector field v modelling displacements on I. On the other hand, large-deformation flow models such as LDDMM traditionally have not integrated random variations directly into the dynamics (see for example [5] for another stochastic approach of diffeomorphism flows). Natural families of probability distributions and generative models taking values in nonlinear spaces such as deformation spaces are generally non-trivial to construct. However, the model proposed in this paper achieves exactly that.

A direct metamorphosis equivalent of the mixed-effects model (4.2) has $\eta = \eta_0$ the population average η, sets $u_0 = \nu_0 = 0$ and encodes the random effects ϕ and ν in (4.2) in the stochastic increments du_t and $d\nu_t$. The action of g_t on the signal is the right action $g_t.f = f \circ g_t^{-1}$ as in (4.2). Now $d\nu_t$ models pure amplitude variation, du_t phase variation, and the combined stochastic evolution of the signal is $df_t = dn_t$. We then assume the observed signal is $f = f_T$ for a fixed end time T of the stochastic process. Spatial correlation in both the deformation increments du_t and the amplitude increments $d\nu_t$ is encoded in the fields σ_l^u and σ_k^ν respectively.

In the above model, the template is stationary in time when disregarding the stochasticity. However, allowing non-zero initial momenta u_0 and ν_0 in both phase and amplitude allows the template to vary with time and thereby gives a nonlinear generalization of a standard multivariate regression model with one latent variable for phase and one for amplitude. This in particular allows modelling of trends over populations where subjects are affected by both the population trend and individual stochastic perturbations.

4.3. Statistical nonlinear modelling

It may initially seem overly complicated to use the metamorphosis framework for a simple regression model. However, statistical models that in linear space seem completely standard are often inherently difficult to generalize to nonlinear spaces. In general, the lack of vector space structure makes distributions and generative models hard to specify, see e.g. [28, 29] for examples of the geometric complexities of generalizing the Euclidean normal distribution.

In Euclidean space, random vectors can model random perturbations. In nonlinear spaces, the lack of a vector space structure prevents this and random perturbations are often most naturally expressed with sequences of infinitesimal steps. Vectors are thus replaced with tangent bundle valued sequences that, when integrated over time, give rise to stochastic flows. When modelling both deterministic and random variations, stochasticity generally couples non-trivially with the deterministic evolution. In addition, perturbations and correlation structure must be specified with respect to a frame of reference. While Euclidean space provides a global coordinate system allowing this, a model of transport must be specified in nonlinear spaces. The stochastic metamorphosis model is an example of a model coupling deterministic and stochastic evolution and using right-invariance to provide reference frames for the perturbations and correlation structure. An example of a related but different approach is [18] where parallel transport is used to link covariance between tangent spaces.

Acknowledgments

AA acknowledges EPSRC funding through award EP/N014529/1 via the EPSRC Centre for Mathematics of Precision Healthcare. AA and DH are partially supported by the European Research Council Advanced Grant 267382 FCCA held by DH. DH is also grateful for partial support from EPSRC Grant EP/N023781/1. SS is partially supported by the CSGB Centre for Stochastic Geometry and Advanced Bioimaging funded by a grant from the Villum Foundation. The authors would like to thank the Isaac Newton Institute for Mathematical Sciences for their support and hospitality during the programme *Growth Form and Self-organisation* when this paper was finished. This program is supported by EPSRC Grant Number EP/K032208/1.

References

1. S. Allassonnière, Y. Amit, and A. Trouvé. Towards a coherent statistical framework for dense deformable template estimation. *Journal of the Royal Statistical Society: Series B (Statistical Methodology)*, 69(1):3–29, 2007.
2. A. Arnaudon, A. L. Castro, and D. D. Holm. Noise and dissipation on coadjoint orbits. *Journal of Nonlinear Science*, 28(1):91–145, 2018.

3. A. Arnaudon, D. D. Holm, and S. Sommer. A geometric framework for stochastic shape analysis. *Foundations of Computational Mathematics*, 19(3):653–701, 2019.

4. M. F. Beg, M. I. Miller, A. Trouvé, and L. Younes. Computing large deformation metric mappings via geodesic flows of diffeomorphisms. *International Journal of Computer Vision*, 61(2):139–157, 2005.

5. A. Budhiraja, P. Dupuis, V. Maroulas, et al. Large deviations for stochastic flows of diffeomorphisms. *Bernoulli*, 16(1):234–257, 2010.

6. B. Charlier, N. Charon, and A. Trouvé. The Fshape Framework for the Variability Analysis of Functional Shapes. *Foundations of Computational Mathematics*, 17(2):287–357, 2017.

7. M. Chen, Y. Zhang, et al. A two-component generalization of the Camassa-Holm equation and its solutions. *Letters in Mathematical Physics*, 75(1):1–15, 2006.

8. G. E. Christensen, R. Rabbitt, and M. I. Miller. Deformable templates using large deformation kinematics. *Image Processing, IEEE Transactions on*, 5(10), 1996.

9. P. Dupuis, U. Grenander, and M. I. Miller. Variational Problems on Flows of Diffeomorphisms for Image Matching. *Quarterly of Applied Mathematics*, 1998.

10. P. T. Fletcher and M. Zhang. Probabilistic Geodesic Models for Regression and Dimensionality Reduction on Riemannian Manifolds. In *Riemannian Computing in Computer Vision*, pages 101–121. Springer, Cham, 2016.

11. F. Gay-Balmaz and T. S. Ratiu. The geometric structure of complex fluids. *Advances in Applied Mathematics*, 42(2):176–275, 2009.

12. D. D. Holm. Euler-Poincaré dynamics of perfect complex fluids. *Geometry, Mechanics, and Dynamics*, pages 169–180, 2002.

13. D. D. Holm. Variational principles for stochastic fluid dynamics. *Proceedings of the Royal Society of London A: Mathematical, Physical and Engineering Sciences*, 471(2176):20140963, 2015.

14. D. D. Holm. Stochastic metamorphosis in imaging science. *Annals of Mathematical Sciences and Applications*, 3(1):309–335, 2018.

15. D. D. Holm, A. Trouvé, and L. Younes. The Euler-Poincaré theory of metamorphosis. *Quarterly of Applied Mathematics*, 67(4):661–685, 2009.

16. D. D. Holm and T. M. Tyranowski. Variational principles for stochastic soliton dynamics. In *Proc. R. Soc. A*, volume 472, page 20150827. The Royal Society, 2016.

17. S. Joshi, B. Davis, M. Jomier, and G. Gerig. Unbiased diffeomorphic atlas construction for computational anatomy. *NeuroImage*, 23:151–160, 2004.

18. L. Kühnel and S. Sommer. Stochastic Development Regression on Non-linear Manifolds. In *Information Processing in Medical Imaging*, Lecture Notes in Computer Science, pages 53–64. Springer, Cham, June 2017.

19. L. Kühnel, S. Sommer, A. Pai, and L. Raket. Most Likely Separation of Intensity and Warping Effects in Image Registration. *SIAM Journal on Imaging Sciences*, 10(2):578–601, 2017.

20. J. S. Marron, J. O. Ramsay, L. M. Sangalli, and A. Srivastava. Func-

tional Data Analysis of Amplitude and Phase Variation. *Statistical Science*, 30(4):468–484, 2015.

21. S. Marsland and T. Shardlow. Langevin equations for landmark image registration with uncertainty. *SIAM Journal on Imaging Sciences*, 10(2):782–807, 2017.

22. M. I. Miller and L. Younes. Group actions, homeomorphisms, and matching: A general framework. *International Journal of Computer Vision*, 41(1-2):61–84, 2001.

23. P. Muralidharan and P. T. Fletcher. Sasaki Metrics for Analysis of Longitudinal Data on Manifolds. *Proceedings / CVPR, IEEE Computer Society Conference on Computer Vision and Pattern Recognition. IEEE Computer Society Conference on Computer Vision and Pattern Recognition*, 2012:1027–1034, 2012.

24. M. Niethammer, Y. Huang, and F.-X. Vialard. Geodesic Regression for Image Time-Series. In *Medical Image Computing and Computer-Assisted Intervention - MICCAI 2011*, Lecture Notes in Computer Science, pages 655–662. Springer, Berlin, Heidelberg, 2011.

25. B. Øksendal. *Stochastic Differential Equations: An Introduction with Applications*. Springer Science & Business Media, 2003.

26. A. Pai, S. Sommer, L. L. Raket, L. Kühnel, S. Darkner, L. Sørensen, and M. Nielsen. A Statistical Model for Simultaneous Template Estimation, Bias Correction, and Registration of 3d Brain Images. 2016.

27. L. L. Raket, S. Sommer, and B. Markussen. A nonlinear mixed-effects model for simultaneous smoothing and registration of functional data. *Pattern Recognition Letters*, 38:1–7, 2014.

28. S. Sommer. Anisotropic Distributions on Manifolds: Template Estimation and Most Probable Paths. In *Information Processing in Medical Imaging*, volume 9123 of *Lecture Notes in Computer Science*, pages 193–204. Springer Berlin Heidelberg, 2015.

29. S. Sommer and A. M. Svane. Modelling anisotropic covariance using stochastic development and sub-Riemannian frame bundle geometry. *Journal of Geometric Mechanics*, 9(3):391–410, 2017.

30. A. Trouvé. An infinite dimensional group approach for physics based models in pattern recognition. *preprint*, 1995.

31. A. Trouvé and F.-X. Vialard. Shape splines and stochastic shape evolutions: a second order point of view. *Quarterly of Applied Mathematics*, 70(2):219–251, 2012.

32. A. Trouvé and L. Younes. Local geometry of deformable templates. *SIAM Journal on Mathematical Analysis*, 37(1):17–59, 2005.

33. A. Trouvé and L. Younes. Metamorphoses through Lie group action. *Foundations of Computational Mathematics*, 5(2):173–198, 2005.

34. J. D. Tucker, W. Wu, and A. Srivastava. Generative models for functional data using phase and amplitude separation. *Computational Statistics & Data Analysis*, 61(Supplement C):50–66, 2013.

35. F.-X. Vialard. Extension to infinite dimensions of a stochastic second-order

model associated with shape splines. *Stochastic Processes and their Applications*, 123(6):2110–2157, 2013.

36. L. Younes, F. Arrate, and M. I. Miller. Evolutions equations in computational anatomy. *NeuroImage*, 45(1, Supplement 1):S40–S50, 2009.

37. M. Zhang, N. Singh, and P. T. Fletcher. Bayesian estimation of regularization and atlas building in diffeomorphic image registration. In *International Conference on Information Processing in Medical Imaging*, pages 37–48. Springer, 2013.

Piecewise Rigid Motion in Diffeomorphism Groups with Strong Right-Invariant Metrics

Dai-Ni Hsieh[*] and Laurent Younes[†]

*Department of Applied Mathematics and Statistics and
Center for Imaging Science, Johns Hopkins University, USA*
[*]dnhsieh@jhu.edu
[†]laurent.younes@jhu.edu

The Large Deformation Diffeomorphic Metric Mapping (LDDMM) algorithms rely on a sub-Riemannian metric on the diffeomorphism group with strong smoothness requirements. This paper will describe a series of simple simulations illustrating the effect of such metrics when considering the motion of rigid objects subject to the associated least action principle. The construction results in an optimal control problem where the control belongs to a product of Euclidean groups. The associated geodesic equations show a variety of behaviors, that are illustrated with experiments, involving various shape configurations. We will explore, in particular, the impact of boundary conditions on the motion, and the effect of repelling forces between shapes, obtained through the addition of a potential energy to the model.

1. Introduction

Shape (or image) registration has become a routine step in the data processing pipeline of computational anatomy [11, 27, 13, 37, 35, 5, 17, 39, 21, 18, 6]. These methods, that align target (subject) images to a reference (or template), can be seen as providing a representation of these images in the diffeomorphism group, i.e., they represent the targets in a coordinate system, where each of these coordinates is a diffeomorphism.

Large deformation diffeomorphic metric mapping, or LDDMM [19, 20, 7, 8, 9, 15, 10, 29], is a family of registration algorithms in which the diffeomorphic representation can be associated with a coordinate system in a Riemannian manifold, providing additional structure that can be leveraged to design various analysis methods, based, for example, on tangent space representations [38, 26, 40], such as regression [12, 14], splines [36] and others for data in infinite-dimensional shape spaces.

LDDMM is typically implemented as an optimal control problem, in which shapes or images are transported (by advection) along the flow of a time-dependent vector field v in the ambient space, which represents the control. The cost associated to this control is, in the original algorithm, defined as a squared norm, $\|v\|_V^2$, in a reproducing kernel Hilbert space (RKHS), V, of sufficiently differentiable vector fields over $\mathbb{R}^d$ where $d = 2$ or 3. More recently, several modifications have been introduced in order to modify this cost, making it, in particular, dependent on the shape that is being advected, leading, for example, to models in which shape-dependent constraints are imposed [42, 1, 16, 2], or using norms relying on the geometry of the deformed objects [43]. In this paper, we explore the very special case in which one advects several shapes subject to the constraint that the restriction of the vector field v to each of them coincides with a Euclidean transformation. This results on a shape space of configurations of rigid objects, on which we will describe the equations of motion, that we will illustrate with numerical experiments.

We describe the related variational problem, and some reductions, in section 2, prove the existence of solutions in section 3 and develop its optimality conditions in section 4. We illustrate this with experiments on the boundary-value problem (section 5) and on the initial-value problem (section 6) in the two-dimensional case. We conclude the paper with a discussion of some open problems in section 7.

2. Problem Formulation

We consider a shape space $\mathcal{M}$ whose elements are one-to-one functions $\gamma : E \to \mathbb{R}^d$, assuming, unless otherwise specified, that E is a finite set. With this assumption, $\mathcal{M}$ is an open subset of the finite-dimensional space $\mathcal{Q}$ containing all functions from E to $\mathbb{R}^d$. We will assume that a partition of E, denoted $(E_1, \ldots, E_m)$ is given and denote by γ_k the restriction $\gamma_{|E_k}$. Using this partition, we will interpret elements of $\mathcal{M}$ as non-overlapping, labelled, families of distinct labelled points in $\mathbb{R}^d$.

Let V be an RKHS of vector fields in $\mathbb{R}^k$ continuously embedded in $C_0^p(\mathbb{R}^d, \mathbb{R}^d)$. Recall that to such spaces is associated a kernel $K_V : \mathbb{R}^d \to \mathbb{R}^d \times M_d(\mathbb{R})$ (the space of $d \times d$ real matrices), such that the associated operator, also denoted K_V, and defined by

$$(K_V f)(x) = \int_{\mathbb{R}^d} K_V(x, y) f(y) \, dy$$

for compactly supported measurable functions $f : \mathbb{R}^d \to \mathbb{R}^d$ can be extended as an operator from the dual space V^* to V, with the property that

$$\mu(v) = \langle K_V \mu, v \rangle_V$$

for all $\mu \in V^*$, $v \in V$. We will assume, to simplify, that K_V is "scalar", i.e., that it takes the form $K_V = K \operatorname{Id}_d$ (where Id_d is the identity matrix on $\mathbb{R}^d$). Here $K : \mathbb{R}^d \times \mathbb{R}^d \to \mathbb{R}$ is a scalar reproducing kernel, and V is typically chosen so that K has an explicit simple form. Well-known kernels include Gaussian kernels

$$K(x,y) = e^{-\frac{|x-y|^2}{2\sigma^2}} \tag{2.1}$$

and Sobolev kernels

$$K(x,y) = P_c \left(\frac{|x-y|}{\sigma} \right) e^{-\frac{|x-y|}{\sigma}}, \tag{2.2}$$

where P_c is a *reverse Bessel polynomial* of degree c (see [22]), normalized so that $P_c(0) = 1$. In the following, K will always be assumed to be radial, i.e., a function of $|x-y|$, like in these examples. Also, for reasons that will become clear in section 7, our experiments will mostly use Sobolev kernels in this paper, and will refer to the parameter σ as the width of the kernel, and to c as its order.

Returning to γ, we define the subspace V_γ containing the vector fields $v \in V$ such that, for all $k = 1, \ldots, m$, there exists A_k, a d by d skew-symmetric matrix, and $\tau_k \in \mathbb{R}^d$, such that $v(\gamma(x)) = A_k \gamma(x) + \tau_k$ for all $x \in E_k$. We then consider the optimal control problem minimizing

$$\frac{1}{2} \int_0^1 \|v(t)\|_V^2 \, dt + U(\gamma(1, \cdot)) \tag{2.3}$$

subject to $\gamma(0, \cdot) = \gamma_0$, $\partial_t \gamma(t, \cdot) = v(t, \gamma(t, \cdot))$ and $v(t) \in V_{\gamma(t)}$ for all t. In other terms, we are using an "LDDMM energy" with the constraint that each subset $\gamma(E_k)$ follows a rigid (Euclidean) motion, and will therefore refer to this class of problems as "piecewise Euclidean LDDMM".

Let G denote the group $(SO_d \ltimes \mathbb{R}^d)^m$ and $\mathfrak{g} = (\mathfrak{so}_d \times \mathbb{R}^d)^m$ its Lie algebra, where $\mathfrak{so}_d$ is the space of d by d skew symmetric matrices. Introduce the control $\theta = (A_1, \tau_1, \ldots, A_m, \tau_m) \in \mathfrak{g}$. Define $\theta \cdot \gamma : E \to \mathbb{R}^d$ by

$$(\theta \cdot \gamma)(x) = A_k \gamma(x) + \tau_k, \quad x \in E_k, \quad k = 1, \ldots, m.$$

Then, the previous problem can clearly be reduced to minimizing

$$\frac{1}{2} \int_0^1 \|\theta(t)\|_{\gamma(t)}^2 \, dt + U(\gamma(1)) \tag{2.4}$$

subject to $\gamma(0) = \gamma_0$, $\partial_t \gamma(t) = \theta(t) \cdot \gamma(t)$, with

$$\|\theta\|_\gamma = \min\{\|v\|_V : v \circ \gamma = \theta \cdot \gamma\}.$$

This norm can be made explicit using the reproducing kernel $K_V = K \operatorname{Id}_{\mathbb{R}^d}$ of V. Introduce the linear operator K_γ defined over functions $\alpha : E \mapsto \mathbb{R}^d$ by

$$K_\gamma : \alpha(\cdot) \mapsto \sum_{x \in E} K(\gamma(\cdot), \gamma(x)) \alpha(x).$$

Then, (using standard results about interpolation in RKHS's [4]), the minimum in the definition of $\|\theta\|_\gamma$ is achieved for

$$v(\cdot) = \sum_{x \in E} K(\cdot, \gamma(x)) \alpha(x)$$

with

$$\alpha = K_\gamma^{-1}(\theta \cdot \gamma).$$

One has moreover

$$\|\theta\|_\gamma^2 = \left(\theta \cdot \gamma \,\middle|\, K_\gamma^{-1}(\theta \cdot \gamma) \right)$$

with the notation

$$(p \,|\, u) = \sum_{x \in E} p(x)^T u(x).$$

Notice that $(\theta, \gamma) \mapsto \|\theta\|_\gamma^2$ is a smooth function of $(\theta, \gamma) \in \mathfrak{g} \times \mathcal{M}$.

From a computational viewpoint, assuming that the elements of E are ordered so that $E = \{x_1, \ldots, x_N\}$ for some integer N, and representing functions α as Nd-dimensional column vectors stacking the vectors $\alpha(x_1), \ldots, \alpha(x_d)$, one can identify, in the same way, $K_\gamma \alpha$ to the Nd-dimensional vector

$$(\tilde{K}_\gamma \otimes \operatorname{Id}_d) \alpha$$

where $\tilde{K}_\gamma$ is the $N \times N$ matrix with entries $K(\gamma(x_k), \gamma(x_l))$, $k, l = 1, \ldots, N$ and $\otimes$ refers to the Kronecker product.

3. Existence of Solutions

Theorem 1. *Assume that U is a continuous function on $\mathcal{M}$. Then there exists at least one minimizer $v \in L^2([0,1], V)$ of variational problem (2.3).*

Elements $v \in V_\gamma$ are characterized by the conditions

$$
\begin{cases}
Dv(\gamma(x)) + Dv(\gamma(x))^T = 0, & x \in E, \\
Dv(\gamma(x)) = Dv(\gamma(y)), & x, y \in E_k, \quad k = 1, \ldots, m, \\
v(\gamma(x)) - Dv(\gamma(x))\gamma(x) = v(\gamma(y)) - Dv(\gamma(y))\gamma(y), & x, y \in E_k, \quad k = 1, \ldots, m,
\end{cases}
$$

which can be written in the form $C_\gamma v = 0$ for an operator C_γ taking values in the finite-dimensional space of functions from $E \times E$ to $\mathbb{R}^q$ for some positive q. Because V is continuously embedded in $C_0^p(\mathbb{R}^d, \mathbb{R}^d)$, with $p \geq 1$, these constraints are continuous in γ and weakly continuous in v, and Theorem 1 in [1] can be applied to prove the existence of a minimizer.

4. Equations of Motion

The Pontryagin maximum principle [41, 23] implies that there exists a time-dependent "co-state" $p : E \to \mathbb{R}^d$, such that the following equations are satisfied:

$$
\begin{cases}
\partial_t \gamma = \theta \cdot \gamma \\
\partial_t p = -\partial_\gamma H_\theta(p, \gamma) \\
\partial_\theta H_\theta(p, \gamma) = 0
\end{cases}
\tag{4.1}
$$

where

$$
H_\theta(p, \gamma) = (p \mid \theta \cdot \gamma) - \frac{1}{2}(\theta \cdot \gamma \mid K_\gamma^{-1}(\theta \cdot \gamma)).
\tag{4.2}
$$

Introduce the operator

$$
\xi_\gamma : \theta \mapsto \xi_\gamma \theta = \theta \cdot \gamma,
$$

and, for $\theta = (A_1, \tau_1, \ldots, A_m, \tau_m) \in \mathfrak{g}$, the operator η_θ defined over functions $u : E \to \mathbb{R}^d$ by

$$
(\eta_\theta u)(x) = A_k u(x), \quad x \in E_k, \quad k = 1, \ldots, m.
$$

Then, the last equation in (4.1) become

$$
\theta = \left(\xi_\gamma^* K_\gamma^{-1} \xi_\gamma \right)^{-1} \xi_\gamma^* p,
$$

and, for $\delta\gamma : E \to \mathbb{R}^d$,

$$
(\partial_\gamma H_\theta \mid \delta\gamma) = (\eta_\theta^* p \mid \delta\gamma) - (\eta_\theta^* \mu \mid \delta\gamma) + \frac{1}{2}(\mu \mid (\partial_\gamma K_\gamma \delta\gamma)\mu)
$$

where $\mu := K_\gamma^{-1}(\theta \cdot \gamma)$. Introduce the notation

$$\left(\partial_\gamma K_\gamma^\dagger(\mu, \mu) \mid \delta\gamma\right) = \left(\mu \mid (\partial_\gamma K_\gamma \delta\gamma)\mu\right).$$

Then, (4.1) can be written as

$$\begin{cases} \partial_t \gamma = \theta \cdot \gamma \\ \partial_t p = -\eta_\theta^*(p - \mu) - \dfrac{1}{2}\partial_\gamma K_\gamma^\dagger(\mu, \mu) \\ \mu = K_\gamma^{-1}(\theta \cdot \gamma) \\ \theta = \left(\xi_\gamma^* K_\gamma^{-1}\xi_\gamma\right)^{-1}\xi_\gamma^* p. \end{cases} \tag{4.3}$$

Remark 1. The original problem can obviously be reduced to a state space $(SO_d \ltimes \mathbb{R}^d)^m$, with state $g = (R_1, T_1, \ldots, R_m, T_m)$, $\gamma = g \cdot \gamma_0$ and $\partial_t g = \theta \cdot g$. The co-state for the new problem is then $(\xi_\gamma^* p)\mathcal{R}_{g^{-1}}^*$, where $\mathcal{R}$ refers to the right translation. We shall not pursue the optimality equations for this reduced problem, but it should come as no surprise, given this remark, that (4.3) can be reduced to a system uniquely specified in terms of $\rho = \xi_\gamma^* p$. Indeed, taking $\alpha \in \mathfrak{g}$, we have

$$\left(\partial_t(\xi_\gamma^* p) \mid \alpha\right) = \left(\partial_t p \mid \xi_\gamma \alpha\right) + \left(p \mid \partial_t(\xi_\gamma \alpha)\right)$$

$$= -\left(p \mid \eta_\theta \xi_\gamma \alpha\right) + \left(\mu \mid \eta_\theta \xi_\gamma \alpha\right) - \frac{1}{2}\left(\mu \mid \partial_\gamma K_\gamma \xi_\gamma \alpha)\mu\right) + \left(p \mid \eta_\alpha \xi_\gamma \theta\right).$$

Noting that $\eta_\theta \xi_\gamma \alpha - \eta_\alpha \xi_\gamma \theta = \xi_\gamma[\theta, \alpha] = \xi_\gamma \mathrm{ad}_\theta \alpha$, we can rewrite (4.3) in the form

$$\begin{cases} \partial_t \gamma = \theta \cdot \gamma \\ \partial_t \rho = -\mathrm{ad}_\theta^* \rho + \xi_\gamma^* \eta_\theta^* \mu - \dfrac{1}{2}\xi_\gamma^* \partial_\gamma K_\gamma^\dagger(\mu, \mu) \\ \mu = K_\gamma^{-1}(\theta \cdot \gamma) \\ \theta = \left(\xi_\gamma^* K_\gamma^{-1}\xi_\gamma\right)^{-1} \rho. \end{cases} \tag{4.4}$$

Remark 2. It is worth mentioning that in the case $m = 1$ (single shape), the equation of motion become equivalent to standard solid body motion, provided that the inertia operator is defined as

$$\mathcal{I}_\gamma(A, A) = \sum_{x,y \in E} G_\gamma(x, y)(A(\gamma(x) - c_\gamma))^T (A(\gamma(y) - c_\gamma))$$

for any skew symmetric matrix A, with $G_\gamma = K_\gamma^{-1}$ and where c_γ is the "center of mass" defined by

$$c_\gamma = \frac{\sum_{x,y \in E} G_\gamma(x,y)\gamma(x)}{\sum_{x,y \in G} G_\gamma(x,y)}. \tag{4.5}$$

In particular, c_γ evolve along a straight line for optimal trajectories, and the body rotation speed around its center of mass is constant. The only factor that remains metric-dependent in this motion is therefore the position of the center of mass. (Obviously, this remark only holds with $m = 1$.)

In the following sections, we provide some experimental illustrations of the properties of this system of equations when applied to some configurations of interest. It is known, from numerical experiments in [24, 25], for example, that geodesics may have interestingly complex patterns when landmark configurations are considered (i.e., when each E_k is a singleton). We will retrieve similar patterns in our case and others specific to the introduction of the rotation group in the model. We will also develop some variations of the basic model, creating, in particular, repelling effects between subsets, providing additional features in the behavior of solutions. We will also describe the impact of Dirichlet boundary conditions on the optimal trajectories, that we will approximate via the introduction of a new finite set, E_0 on which the velocity v is prescribed to vanish. Adding this condition to the definition of the space, V_γ, one obtains similar equations of motion, simply replacing $\theta \cdot \gamma$ by $(0, \ldots, 0, \theta \cdot \gamma)$ with as many d-dimensional zeros as elements in E_0.

Our experiments were run in Python on a MacBook laptop (code available from the py-lddmm project on Bitbucket.org).

5. Experiments: Boundary-value Problem (Matching)

In this section, we provide some examples of optimal trajectories for the registration of configurations of shapes. The first rows of figures 1 and 2 provide a confirmation of the remark made for the case $m = 1$, in which one indeed recovers a straight line trajectory for the center of mass. (In this and subsequent figures in this paper, we plot the initial and final shapes of the trajectory, with a green line linking the center of mass of each shape (defined according to equation (4.5)) and some fixed point on the shape. The figures also trace the trajectories of the centers of mass (m of them).)

In the second rows of these two figures, the result of the registration algorithm is also provided with $m = 2$ (the original shapes having two connected components), exhibiting a small variation of the estimated diffeomorphism for a relatively large kernel size, and a more significant one for small kernels.

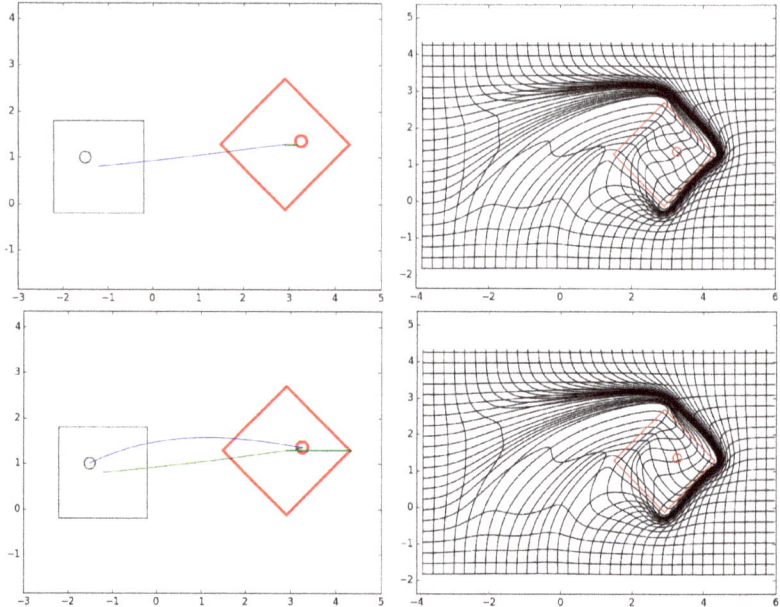

Fig. 1: Rotation of a square with a small inscribed circle via piecewise Euclidean LDDMM, with a Laplacian kernel of width 0.25 and order 3: trajectory of the center(s) of mass and grid deformation. First row: $m = 1$ (single displacement). Second row: $m = 2$ (one displacement per shape). While the two estimated diffeomorphisms are similar, one can note a slight difference inside the square. (Kernel parameters: width = 0.25, order = 3.)

Another matching experiment is depicted in figure 3, with a more complicated problem involving two shapes (a "C" and a circle) that change relative positions and have to avoid each other. Finally, figure 4 provides an experiment with a Dirichlet boundary condition, in which a vertical line (depicted in orange) remains fixed, forcing the shape to take an avoiding trajectory.

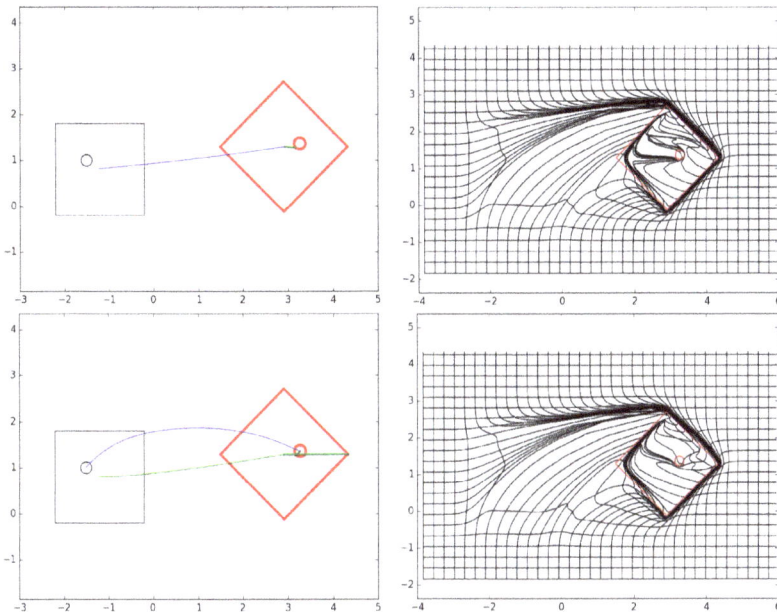

Fig. 2: Same as Figure 1, with a kernel width of 0.1. The difference between the diffeomorphisms within the square is now more significant.

6. Experiments: Initial Value Problem (Shooting)

We now provide some illustrations of geodesic paths from the shooting viewpoint, in which the optimality equations (4.3) or (4.4) are solved with some initial shapes $\gamma(0, \cdot)$ and control $\theta(0)$. When solving (4.3), the initial co-state, $p(0)$ is taken equal to $\mu(0) = K_{\gamma(0)}^{-1} \xi_{\gamma(0)} \theta(0)$, since the maximum principle implies

$$\xi_{\gamma(0)}^* \mu(0) = \xi_{\gamma(0)}^* K_{\gamma(0)}^{-1} \xi_{\gamma(0)} \theta(0) = \xi_{\gamma(0)}^* p(0)$$

and we have seen that the solution of the equation only depended on $\xi_{\gamma(0)}^* p(0)$.

Figure 5 provides geodesic trajectories for a pair of circles shot in parallel, opposite directions. Similar to the landmark case [24, 25], the circles are attracted to each other and, depending on the distance between the initial parallel lines, their trajectories may either be bounded, both shapes rotating around a limit attractor point, or be unbounded, each shape finally avoiding each other. Trajectories near the transition point, such as the one shown in the first panel of figure 5, show the shapes turning around

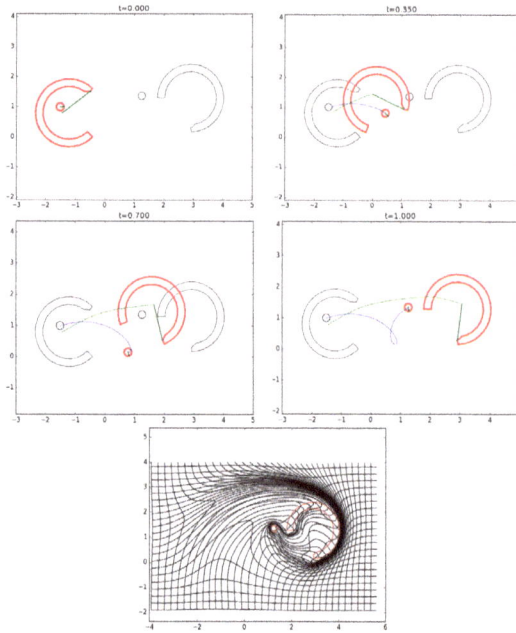

Fig. 3: First two rows: Four time-points of the optimal motion of a "C" shape and a small circle subject to different displacements in which they change position. The circle's trajectory makes a loop while making way for the "C". Third row: Estimated diffeomorphism.

each other for a while before separating. The next two panels in this figure show that angular momenta modify the shape trajectories, resulting, in this example, in almost no attraction when the shapes turn in the counterclockwise direction and a rapid attraction in the clockwise direction.

We next illustrate the impact of a boundary condition on the motion of a single shape initialized with non-zero linear velocity $\tau_1(0)$ and angular velocity $A_1(0)$. Without any condition, the motion is, as expected, a translation of the center of mass combined with a rotation of the object around it. When the shape is inscribed in a box that is constrained to remain invariant, the center of mass trajectory now has curvature and turns in the opposite direction of the initial shape rotation.

Fixed shapes can also have an affect similar to an attraction force on moving shapes that pass nearby, to prevent too large deformations that such events may produce. An illustration is provided in figure 7. Figure 8

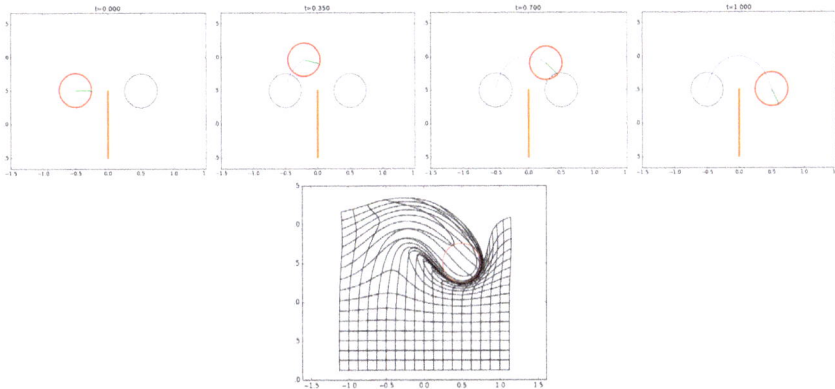

Fig. 4: First row: Four time-points of the optimal motion of a circle avoiding a fixed vertical line. Second row: Estimated diffeomorphism.

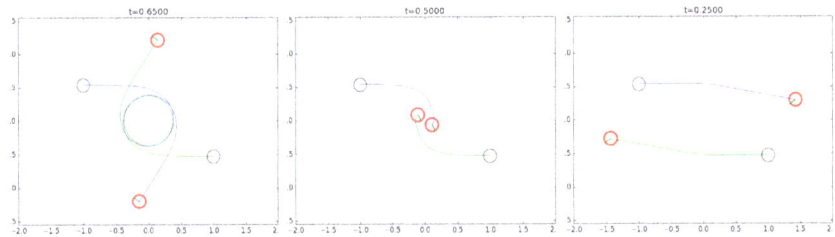

Fig. 5: Geodesic evolution of two circles initialized with parallel and opposite linear momenta. Left: No angular momentum: the circles turn around each other before separating again. Center: Clockwise angular momentum: the shapes are rapidly attracted to each other. Right: Counter-clockwise angular momentum: the trajectories are slightly deflected but the shapes remain far away from each other.

provides another illustration of this effect, this time with multiple shapes that tend to cluster during the motion.

We conclude this section with some experiments of a modified model in which one adds a repelling potential energy to the Hamiltonian, therefore replacing (6.1) by

$$H_\theta(p, \gamma) = (p \mid \theta \cdot \gamma) - \frac{1}{2}\big(\theta \cdot \gamma \mid K_\gamma^{-1}(\theta \cdot \gamma)\big) + U(\gamma). \qquad (6.1)$$

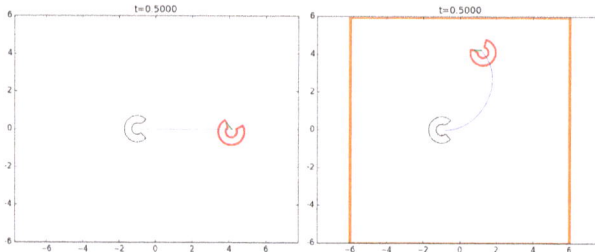

Fig. 6: Comparison of the geodesic evolution of a rigid shape without (left) and with (right) boundary condition. In the right panel, the fixed boundary is a box represented in orange.

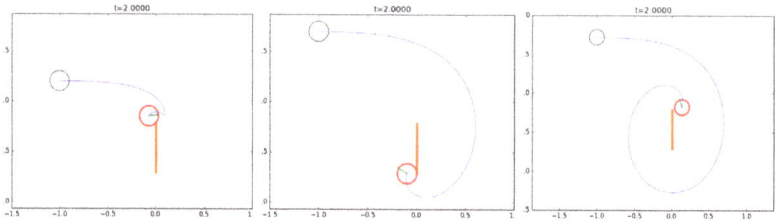

Fig. 7: Three illustration of the attraction effect exerted by a fixed line on a circle passing in its vicinity. (Kernel width: 0.5, order: 3.)

Here, we take

$$U(\gamma) = \sum_{\substack{k,l=1 \\ k \neq l}}^{m} \sum_{x \in E_k} \sum_{y \in E_l} \frac{c}{|\gamma(x) - \gamma(y)|^a}$$

for some positive constants c and a. After adding this term, new equations of motion are obtained by subtracting $\partial_\gamma U$ (resp. $\xi_\gamma^* \partial_\gamma U$) to the right-hand side of the second equation of system (4.3) (resp. (4.4)).

Two solutions of the resulting system are depicted in figure 9. As expected, they show that the repelling force of the new energy term progressively balances the attraction effect resulting from the diffeomorphic energy before taking over, resulting in trajectories that arguably offer more variety than the one associated with "plain LDDMM". It is important, however, to point out that adding a potential energy invalidates the Riemannian (or sub-Riemannian) interpretation of the model, and the optimal trajectories deriving from Hamilton's principle are not geodesics anymore (they are stationary points of the energy, not necessarily local minimizers). As a result, many of the important properties of the Riemannian structure, such as the

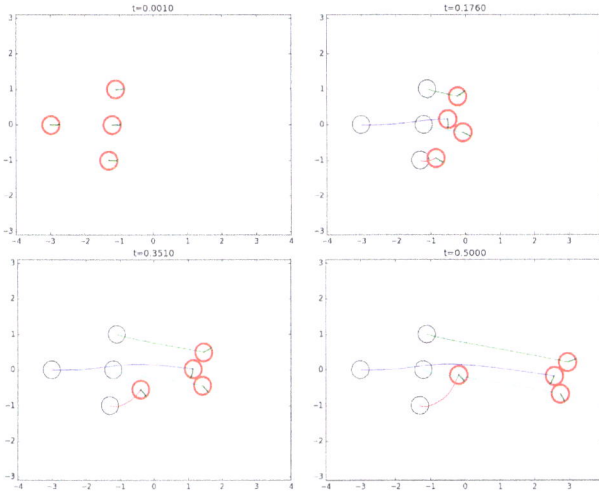

Fig. 8: Four time-points of a geodesic involving four circles. At $t = 0$, only the left-most circle has non-zero linear and angular velocities. Two other circles cluster with it along the trajectory, while the fourth one is left behind.

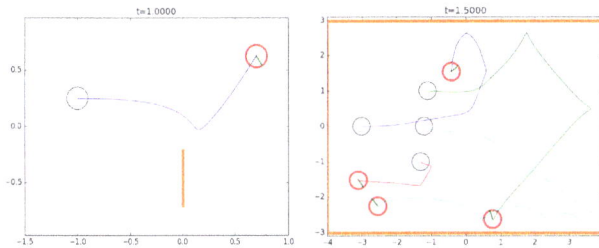

Fig. 9: Two examples of shape trajectories with repelling potential energy. Left: Same example as in figure 7, with a circle now rebounding on the fixed line after being attracted by it (kernel width: 0.5). Right: Multiple circles inscribed in a fixed rectangle, with all initial linear and angular velocity vanishing except those of the leftmost circle (kernel width: 0.2).

associated distance, or exponential map representations, are lost. This construction may be, however, of interest when defining mechanistic models of shape evolution, either deterministic or stochastic [32, 33, 30, 3, 31], for which purely geodesic motion may be too limited.

7. Discussion

We have limited our discussion so far to finite sets of points, but a natural issue is whether an extension to infinite sets (e.g., smooth curves) would be valid. This raises several difficulties, however, the first one being whether the sets V_γ of piecewise Euclidean velocities are non-trivial, i.e., is $V_\gamma \neq \{0\}$? In other terms, given, a function $\gamma : E \to \mathbb{R}^d$, with E possibly infinite, and given $\theta \in (\mathfrak{so}_d(\mathbb{R}) \ltimes \mathbb{R}^d)^m$, does there exist $v \in V$ such that $v(\gamma(x)) = \theta \cdot \gamma(x)$?

We obviously want to assume enough smoothness on these functions, and restrict, to simplify, to the case when $E = E_1 \cup \cdots \cup E_m$ is a union of disjoint smooth compact manifolds and γ is a C^∞ embedding. Then it should be clear that, as soon as the sets $\gamma(E_k)$ are disjoint, a sufficient condition for the existence of such $v \in V$ is that V contains all C^∞ functions with compact support. (Because it then suffices to define v separately in small neighborhoods of each shape.) This includes, in particular, all the RKHS's V that are equivalent to Sobolev, such as those we have used in our experiments. In such cases, the proof of the existence of solutions to the boundary-value problem (section 3) is valid with almost no modification.

The problem is more difficult, and mainly open, if one considers Gaussian RKHS's, since the previous argument does not apply. It is known, (see, for example, [34, 28]) that if $\gamma(E_k)$ has a non-empty interior in $\mathbb{R}^d$, then there are no non-vanishing $v \in V$ of which the restriction to $\gamma(E_k)$ is polynomial. This automatically implies that there can be no non-constant function on $\gamma(E_k)$, and no linear invertible function either, as soon as this set contains a linear segment with non-empty relative interior.

On the other hand, if $m = 1$ and $\gamma(E)$ is included in a sphere, then V_γ is indeed non-trivial [34, 28], so that the previous situation is not universal. Deciding, given an arbitrary collection of curves, whether $V_\gamma = \{0\}$ or not remains an open problem, up to our knowledge.

Note that the operator K_γ defined in section 2 is still well defined in the general case, because it is the kernel operator of the RKHS

$$V^{(\gamma)} = \{f = v_{|\gamma(E)} : v \in V\}$$

formed with the restrictions to $\gamma(E)$ of vector fields in V, with the norm

$$\|f\|_{V^{(\gamma)}} = \min\{\|v\|_V : f = v_{|\gamma(E)}\}.$$

This implies that the equations of motion determined in section 4 remain meaningful, as soon as $\theta \cdot \gamma \in V^{(\gamma)}$, which is the issue that we just discussed.

There are, however, significant numerical issues attached to the implementation of these equations, which can be observed already in the discrete case when the discretization step of the curves is small compared to the kernel width. In such cases, K_γ (considered as a matrix) is ill-conditioned, with most eigenvalues below computer precision, and the computation of $K_\gamma^{-1}(\theta \cdot \gamma)$ is delicate. In our implementation, we project $\theta \cdot \gamma$ on the eigenspaces associated to large enough eigenvalues of K_γ before solving the linear system, which has some stabilizing effect, even though this does not seem to completely solve the issue. The same numerical problems also arise when shapes gravitate too close to each other. It is important to notice that these issues are not specific to the situation that we consider here (piecewise Euclidean LDDMM), but arise with general registration problems for curves and surfaces involving LDDMM.

The construction that we provided in this paper may be seen as an informative toy model in the framework of free-form shape registration and analysis. It exemplifies, by placing them in the simple context of piecewise Euclidean motion, some of the challenges facing LDDMM when dealing with curves and surfaces, on the theoretical and numerical viewpoints, raising, for example, the issue of devising a numerically stable scheme computing RKHS norms of smooth vector fields along unions of finely discretized curves.

The piecewise Euclidean LDDMM has an interest of its own. It generalizes the widely used landmark-based model to collections of rigid bodies and may be relevant to develop shape analysis studies for datasets in which the relative position of such objects may contain relevant information, such as bone position in a moving body, or spine motion.

References

1. Sylvain Arguillère, Emmanuel Trélat, Alain Trouvé, and Laurent Younes. Shape deformation analysis from the optimal control viewpoint. *Journal de Mathématiques Pures et Appliquées*, 104(1):139–178, 2015.
2. Sylvain Arguillère, Emmanuel Trélat, Alain Trouvé, and Laurent Younes. Registration of multiple shapes using constrained optimal control. *SIAM Journal on Imaging Sciences*, 9(1):344–385, 2016.
3. Alexis Arnaudon, Darryl D. Holm, Akshay Pai, and Stefan Sommer. A stochastic large deformation model for computational anatomy. In *International Conference on Information Processing in Medical Imaging*, pages 571–582. Springer, 2017.
4. N. Aronszajn. Theory of reproducing kernels. *Trans. Am. Math. Soc.*, 68:337–404, 1950.

5. John Ashburner. A fast diffeomorphic image registration algorithm. *NeuroImage*, 38(1):95–113, 2007.
6. John Ashburner and Karl J. Friston. Diffeomorphic registration using geodesic shooting and Gauss–Newton optimisation. *NeuroImage*, 55(3):954–967, 2011.
7. Brian Avants and James C. Gee. Geodesic estimation for large deformation anatomical shape averaging and interpolation. *NeuroImage*, 23:S139–S150, 2004.
8. M. Faisal Beg, Michael I. Miller, Alain Trouvé, and Laurent Younes. Computing large deformation metric mappings via geodesic flows of diffeomorphisms. *International Journal of Computer Vision*, 61(2):139–157, 2005.
9. Yan Cao, Michael I. Miller, Raimond L. Winslow, and Laurent Younes. Large deformation diffeomorphic metric mapping of vector fields. *IEEE Transactions on Medical Imaging*, 24(9):1216–1230, 2005.
10. Can Ceritoglu, Kenichi Oishi, Xin Li, Ming-Chung Chou, Laurent Younes, Marilyn Albert, Constantine Lyketsos, Peter van Zijl, Michael I. Miller, and Susumu Mori. Multi-contrast large deformation diffeomorphic metric mapping for diffusion tensor imaging. *NeuroImage*, 47(2):618–627, 2009.
11. Gary E. Christensen, Richard D. Rabbitt, and Michael I. Miller. Deformable templates using large deformation kinematics. *IEEE Transactions on Image Processing*, 5(10):1435–1447, 1996.
12. B. C. Davis, P. T. Fletcher, E. Bullitt, and S. Joshi. Population Shape Regression From Random Design Data. *2007 IEEE 11th International Conference on Computer Vision*, pages 1–7, 2007.
13. Marc Droske and Martin Rumpf. A variational approach to nonrigid morphological image registration. *SIAM Journal on Applied Mathematics*, 64(2):668–687, 2004.
14. P. Thomas Fletcher. Geodesic regression and the theory of least squares on Riemannian manifolds. *International Journal of Computer Vision*, 105(2):171–185, 2013.
15. Joan Glaunès, Anqi Qiu, Michael I. Miller, and Laurent Younes. Large deformation diffeomorphic metric curve mapping. *International Journal of Computer Vision*, 80(3):317–336, 2008.
16. Barbara Gris, Stanley Durrleman, and Alain Trouvé. A sub-riemannian modular approach for diffeomorphic deformations. In *International Conference on Networked Geometric Science of Information*, pages 39–47. Springer, 2015.
17. Monica Hernandez, Salvador Olmos, and Xavier Pennec. Comparing algorithms for diffeomorphic registration: Stationary lddmm and diffeomorphic demons. In *2nd MICCAI Workshop on Mathematical Foundations of Computational Anatomy*, pages 24–35, 2008.
18. Darryl D. Holm, Alain Trouvé, and Laurent Younes. The Euler-Poincaré theory of metamorphosis. *Quarterly of Applied Mathematics*, 97:661–685, 2009.
19. S. Joshi. *Large Deformation Diffeomorphisms and Gaussian Random Fields for Statistical Characterization of Brain Sub-manifolds*. PhD thesis, Sever institute of technology, Washington University, 1997.
20. Sarang C. Joshi and Michael I. Miller. Landmark matching via large deforma-

tion diffeomorphisms. *IEEE Transactions on Image Processing*, 9:1357–1370, 2000.

21. Arno Klein, Jesper Andersson, Babak A. Ardekani, John Ashburner, Brian Avants, Ming-Chang Chiang, Gary E. Christensen, D. Louis Collins, James Gee, Pierre Hellier, Joo Hyun Song, Mark Jenkinson, Claude Lepage, Daniel Rueckert, Paul Thompson, Tom Vercauteren, Roger P. Woods, J. John Mann, and Ramin V. Parsey. Evaluation of 14 nonlinear deformation algorithms applied to human brain {MRI} registration. *NeuroImage*, 46(3):786–802, 2009.

22. Hi L. Krall and Orrin Frink. A new class of orthogonal polynomials: The Bessel polynomials. *Transactions of the American Mathematical Society*, 65(1):100–115, 1949.

23. Jack Macki and Aaron Strauss. *Introduction to optimal control theory*. Springer Science; Business Media, 2012.

24. Robert I. McLachlan and Stephen Marsland. N-particle dynamics of the Euler equations for planar diffeomorphisms. *Dynamical Systems*, 22(3):269–290, 2007.

25. Mario Micheli, Peter W. Michor, and David Mumford. Sobolev metrics on diffeomorphism groups and the derived geometry of spaces of submanifolds. *Izvestiya: Mathematics*, 77(3):541, 2013.

26. Michael I. Miller, Alain Trouvé, and Laurent Younes. Geodesic shooting for computational anatomy. *Journal of Mathematical Imaging and Vision*, 24(2):209–228, 2006.

27. Michael I. Miller and Laurent Younes. Group actions, homeomorphisms, and matching: A general framework. *International Journal of Computer Vision*, 41(1-2):61–84, 2001.

28. Ha Quang Minh. Some properties of gaussian reproducing kernel hilbert spaces and their implications for function approximation and learning theory. *Constructive Approximation*, 32(2):307–338, 2010.

29. Laurent Risser, F. Vialard, Robin Wolz, Maria Murgasova, Darryl D. Holm, and Daniel Rueckert. Simultaneous multi-scale registration using large deformation diffeomorphic metric mapping. *IEEE Transactions on Medical Imaging*, 30(10):1746–1759, 2011.

30. Tony Shardlow and Stephen Marsland. Langevin equations for landmark image registration with uncertainty. *SIAM Journal on Imaging Sciences*, 10(2):782–807, May 2017.

31. Stefan Sommer, Alexis Arnaudon, Line Kuhnel, and Sarang Joshi. Bridge simulation and metric estimation on landmark manifolds. *arXiv preprint arXiv:1705.10943*, 2017.

32. Valentina Staneva and Laurent Younes. Modeling and estimation of shape deformation for topology-preserving object tracking. *SIAM Journal on Imaging Sciences*, 7(1):427–455, 2014.

33. Valentina Staneva and Laurent Younes. Learning shape trends: Parameter estimation in diffusions on shape manifolds. In *Proceedings of the IEEE Conference on Computer Vision and Pattern Recognition Workshops*, pages 38–46, 2017.

34. Ingo Steinwart, Don Hush, and Clint Scovel. An explicit description of the reproducing kernel Hilbert spaces of Gaussian rbf kernels. *IEEE Transactions on Information Theory*, 52(10):4635–4643, 2006.

35. Martin Styner, Ipek Oguz, Shun Xu, Christian Brechbühler, Dimitrios Pantazis, James J. Levitt, Martha E. Shenton, and Guido Gerig. Framework for the statistical shape analysis of brain structures using spharm-pdm. *The Insight Journal*, 1071:242, 2006.

36. Alain Trouvé and François-Xavier Vialard. Shape Splines and Stochastic Shape Evolutions: A Second Order Point of View. *Quart. Appl. Math.*, 70:219–251, 2012.

37. Alain Trouvé and Laurent Younes. Metamorphoses through Lie group action. *Foundations of Computational Mathematics*, 5(2):173–198, 2005.

38. Marc Vaillant, Michael I. Miller, Laurent Younes, and Alain Trouvé. Statistics on diffeomorphisms via tangent space representations. *NeuroImage*, 23:S161–S169, 2004.

39. Tom Vercauteren, Xavier Pennec, Aymeric Perchant, and Nicholas Ayache. Diffeomorphic demons: Efficient non-parametric image registration. *NeuroImage*, 45(1):S61–S72, 2009.

40. François-Xavier Vialard, Laurent Risser, Daniel Rueckert, and Colin J. Cotter. Diffeomorphic 3D Image Registration via Geodesic Shooting Using an Efficient Adjoint Calculation. *International Journal of Computer Vision*, pages 1–13, 2011.

41. Richard Vinter. *Optimal Control*. Birkaüser, 2000.

42. Laurent Younes. Constrained diffeomorphic shape evolution. *Foundations of Computational Mathematics*, 12(3):295–325, 2012.

43. Laurent Younes. Hybrid Riemannian metrics for diffeomorphic shape registration. *Annals of Mathematical Sciences and Applications*, 2018.

Existence and Continuity of Minimizers for the Estimation of Growth Mapped Evolutions for Current Data Term and Counterexamples for Varifold Data Term

Irène Kaltenmark* and Alain Trouvé†

*INT, UMR 7289, Aix-Marseille Université,
CNRS, Marseille, France
irene.kaltenmark@cmla.ens-cachan.fr

†CMLA, ENS Cachan, CNRS, Université Paris-Saclay,
94235 Cachan, France
alain.trouve@cmla.ens-cachan.fr

In the field of computational anatomy, the complexity of changes occurring during the evolution of a living shape while it is growing, aging or reacting to a disease, calls for more and more accurate models to allow subject comparison. Growth mapped evolutions have been introduced to tackle the loss of homology between two ages of an organism following a growth process that involves creation of new material. They model the evolution of longitudinal shape data with partial mappings. One viewpoint consists in a progressive embedding of the shape into an ambient space on which acts a group of diffeomorphisms. In practice, the shape evolves through a time-varying dynamic called the growth dynamic.

The concept of shape space has now been widely studied and successfully applied to analyze the variability of a population of related shapes. Time-varying dynamics subsequently enlarge this framework and open the door to new optimal control problems for the assimilation of longitudinal shape data. We address in this paper an interesting problem in the field of the calculus of variations to investigate the existence and continuity of solutions for the registration of growth mapped evolutions with the growth dynamic. This theoretical question highlights the unexpected role of the data term grounded either on current or varifold representations. Indeed, in this new framework, the spatial regularity of a continuous scenario estimated from a temporal sequence of shapes with the growth dynamic depends on the temporal regularity of the deformation. Current metrics have the property to be more robust to this spatial regularity than varifold metrics. We will establish the existence and continuity of global minimizers for current data term and highlight two counterexamples for varifold data term.

Contents

1. Introduction

1.1. *Context*

In the field of computational anatomy, the *diffeomorphometry*, as introduced by Miller, Trouvé, and Younes [32, 33], consists in modeling and analyzing the variability of a population of embedded shapes through the action of a group of diffeomorphisms. The viewpoint of optimal control offers an elegant framework for diffeomorphic registration [1, 45, 2]. Initially, the matching between two embedded shapes $\mathcal{S}$ and $\mathcal{S}^{\mathrm{tar}}$ in $\mathbb{R}^d$, like curves or surfaces, consists in the problem

$$\min_{\phi} R(\phi) + \mathcal{A}(\phi) \tag{1.1}$$

where R is a regularization term on the deformation $\phi : \mathbb{R}^d \to \mathbb{R}^d$ and $\mathcal{A}$ measures the discrepancy between the target shape $\mathcal{S}^{\text{tar}}$ and the deformed source shape $\phi(\mathcal{S})$. The construction of diffeomorphism groups with Riemannian metrics allows to address this problem in the setting of high-dimensional deformation space. A group is defined by the space V of vector fields that models its tangent space at the identity. An element of the group is then generated as the end point of a flow $(\phi_t^v)_{t \in [0,1]}$ which satisfies the ordinary differential equation:

$$\dot{\phi}_t = v_t \circ \phi_t, \ \phi_0 = \text{Id} \tag{1.2}$$

where $v_t \in V$ for all t. The existence of geodesics in the group allows to rewrite the variational problem

$$\min_{v} R(v) + \mathcal{A}(\phi_1^v) \tag{1.3}$$

over square integrable vector fields of V and the introduction of Reproducing Kernel Hilbert Spaces (RKHS) to model V leads to efficient numerical methods like the large deformation diffeomorphic metric mapping (LDDMM).

1.2. *Growth mapped evolutions*

With the progress achieved in medical imaging analysis, the interest for longitudinal data set has substantially increased in the last years and requires the processing of complex changes, which especially appear during growth or aging phenomena. Various methods derive from the concept of shape spaces as Riemannian manifolds ranging from parallel transport [37], Riemannian splines [45], geodesic regression [35, 48, 14] including the inference from a population of a prototype scenario of evolution and its spatio-temporal variability [10]. In another context, the unbalanced optimal transport [29, 36] extends the optimal transport framework to tackle the problem of mass creation. Up to now, longitudinal analysis has been limited to the study of data sets with homologous observations. Yet, during the growth or the degeneration of an organism, the changes occurring over time cannot always be explained by diffeomorphic transformations, at least in a biological sense.

In [25], we introduced the concept of *growth mapped evolutions (GME)* that consists in a set of embedded shapes $(S_t)_{t \in T}$ indexed by a time interval $T \subset \mathbb{R}$ and equipped with a flow $(\phi_t)_{t \in T}$ on the embedding space $\mathbb{R}^d$. The constraint of exhaustive homology between any two shapes S_s and S_t is relaxed, as illustrated in Figure 1, by an inclusion condition: for any pair $s \leq t$ in T,

$$\phi_{s,t}(S_s) \subset S_t \,, \tag{1.4}$$

where $\phi_{s,t} = \phi_t \circ \phi_s^{-1}$.

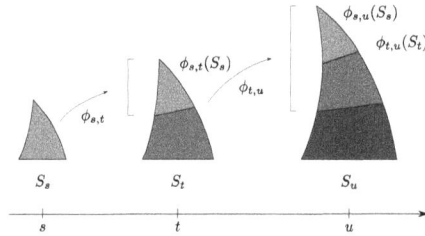

Fig. 1.   Inner partial mappings of a growth mapped evolution to model a creation process.

A growth mapped evolution is therefore a *nested* sequence of all ages of the shape through the flow. The shape S_t is composed partly by the image $\phi_{s,t}(S_s)$ of a previous state and by the new points that appeared between the time s and t calling to consider the time of birth of each point, induced by the colors in the discrete sketch of Figure 1. The viewpoint of coordinate systems with a temporal component allows to describe the history of a creation process and to define more specific populations of GME sharing a common growth pattern. A biological coordinate system consists in a space X called the coordinate space and a function $\tau : X \to T$ called the birth tag whose lower sets generate a collection of nested shapes $X_t = \{x \in X \,|\, \tau(x) \leq t\}$ of X forming a canonical growth scenario. Any individual can then be parametrized with a collection of smooth mappings $q_t : X_t \to \mathbb{R}^d$ so that

$$S_t = q_t(X_t) \,.$$

The level sets $X_{\{t\}} = \{x \in X \,|\, \tau(x) = t\}$ are the new coordinates of X_t created at time t and $q_t(X_{\{t\}})$ models the new points of S_t. We aim to build mappings $(q_t)_t$ that are consistent with the flow, meaning that for any $s < t \in T$ and any $x \in X_s$,

$$q_t(x) = \phi_{s,t}(q_s(x)).$$

If $(\phi_t)_t$ is the flow of a time-varying vector field $v : T \to V$, it follows that

$$\dot{q}_t(x) = v_t(q_t(x)).$$

The mappings $(q_t)_t$ are extended to X to anticipate the appearance of the new points. If the point associated to a coordinate $x \in X$ does not exist at time t, $q_t(x)$ returns its future place of birth. This leads to require that $q_t(x) = q_0(x)$ for any $t \le \tau(x)$ and that

$$\dot{q}_t(x) = \mathbb{1}_{\tau(x) \le t} v_t(q_t(x)). \tag{1.5}$$

This new dynamic that depends both on the control v_t and the time variable t is called the **growth dynamic**. See [24, 23] for more details and for a first study of the following variational problem. Throughout this paper, all scenarios evolve in the fixed time interval $T = [0, 1]$.

1.3. *Contributions*

Consider a population of growth mapped evolutions modeled on a biological coordinate system (X, τ). The estimation of a such growth mapped evolution, given an initial condition q_0 and a target shape $\mathcal{S}^{\text{tar}}$ to reach at the final time of the development, can be expressed as a **minimization problem** on an energy of the type

$$E(v) = \frac{1}{2} \int_0^1 |v_t|_V^2 \, dt + \mathcal{A}(v), \tag{1.6}$$

where v belongs to $L^2([0, 1], V)$, the initial mapping q_0 is fixed, q satisfies the **growth dynamic** 1.5 and $\mathcal{A} : L_V^2 \to \mathbb{R}$ measures the discrepancy between the target and the final shape $q_1(X)$.

In the usual approach of shape spaces, the matchings of two shapes consists in searching a geodesic in a chosen space G_V of diffeomorphisms with constraints on the ends. Retrieving an optimal growth scenario via the growth dynamic does not only constrain the ends of the flow of diffeomorphisms. Indeed, the final status q_1 of a solution cannot be written as an image of the initial mapping q_0 by the final state of the flow ϕ_1^v. It depends instead on the whole evolution of the flow over time and the variational problem 1.3 does not encompass this difficulty. The optimal flow to reach a final target is usually not a geodesic of G_V.

This paper examines the existence and the continuity of global minimizers v to the optimization problem 1.6. In Section 2, we recall that the
spatial regularity of q_1 depends on the temporal regularity of v. The continuity of optimal solutions $v^* \in \mathcal{C}([0,1], V)$ is thus a crucial point. The
usual representations of shapes by currents or varifolds require to ensure
that the shapes are sufficiently regular as they are a priori neither of class
$\mathcal{C}^1$ nor rectifiable. For this purpose, we will exploit in Section 2 the density
of $\mathcal{C}([0,1], V)$ in $L^2([0,1], V)$ as continuous vector fields actually generate
mappings q_1 of $\mathcal{C}^1(X, \mathbb{R}^d)$.

Section 3 presents a rather unexpected result. We exhibit a setting with
a data attachment term built on the representation by varifolds where no
global minimizer is continuous. The difference between representations by
currents or varifolds, regarding the models associated to the growth dynamic, is explained by the fact that oscillations in time of the vector field v
generate oscillations in space for the shapes $q_t(X_t)$. The currents through
their cancellation effect on these spatial oscillations can prevent this
behavior.

In Section 4.1, the existence of solutions in $L^2([0,1], V)$ for a current
data term attachment is shown in a general setting. The proof is based
on the linearity to the tangential component of the current representation
that allows to deduce the lower semi-continuity of $\mathcal{A}$. The continuity of an
optimal vector field v does not follow immediately and will be established
in Section 4.2.

1.4. *Data representation*

As many registration problems in shape analysis, the reconstitution of a
growth mapped evolution requires to use a shape similarity metric. The
seminal work of [47, 17] was the first to exploit the idea from geometric measure theory of representing shapes as elements of a certain space
of distributions leading to distances directly invariant to parametrization.
This approach is based on the representations of oriented curves or surfaces as mathematical currents. Later on, [8] introduced the alternative but
orientation-invariant representation known as varifolds, with a straightforward extension to oriented varifolds [26], before the higher order model of
normal cycles recently investigated in [39]. Another recent approach that
still stems from the representation of shapes by distributions, integrates
unbalanced optimal transport to define a new similarity measure [13].

The focus of this paper is set on currents and varifolds. In the case of currents, the metric is define by a specific kernel linear with respect to the tangential data. This property is well known to make this metric robust to the noise. See Figure 2. Note however, as a downside, that this linearity also prevents the capture of structures like sharp spines or tails. As we will see in this paper, this central difference between current and varifold metrics will have an important impact on how they drive the reconstitution of growth mapped evolutions.

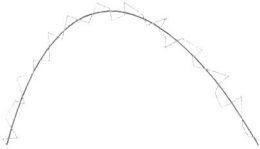

Fig. 2. Denote X the noisy curve and Y the smooth one. From a current point of view, $\mu_X \approx \mu_Y$. Conversely, with varifolds, the length of X is about twice the length of Y and this approximation no longer holds.

2. Extension of the varifold and current representations to foliated shapes

2.1. *Foliation and spatial regularity of the GME's shapes*

Figure 3 illustrates two scenarios generated with the growth dynamic. This time-varying dynamic allows to forecast within the initial mapping q_0 the position of the new points that will progressively be created. In these two examples, all the coordinates are initially set in a horizontal plane of $\mathbb{R}^3$. One can see in Figure 3 that at each time t, the individual shape is displaced upwards and the subset $X_{\{t\}} = \{x \in X \mid \tau(x) = t\}$ of new coordinates is activated (their image model the new points created). When the creation process is regular, e.g. with a progressive addition of regular extensions at one boundary of the shape, the sets of new coordinates $(X_{\{t\}})_t$ define a *foliation* on the coordinate space X. A foliation [38, 28] looks locally like a union of parallel shapes of smaller dimension called the leaves of the foliation. In a general biological coordinate space (X, τ), the existence of a foliation depends on the regularity of the birth tag τ. In this paper, we will consider a canonical situation where $X = [0, 1] \times B$ where B is an oriented compact manifold with corners and τ is just the projection on the first coordinate (for any $(t, b) \in [0, 1] \times B$, $\tau(t, b) = t$). Each leaf $X_{\{t\}} = \{t\} \times B$ is diffeomorphic to B and we denote then $B_t \doteq \{t\} \times B$.

The initial condition q_0 has no reason to be an embedding of the whole coordinate space X. For the first scenario of Figure 3, all leaves of X have the same image in $\mathbb{R}^d$. The shape can be seen as completely folded on itself

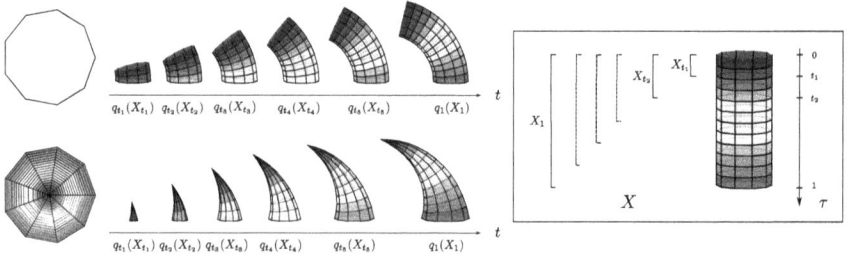

Fig. 3.   Two examples of individual scenarios built on a given biological coordinate system (X, τ). The first column shows the top view of the initial position q_0 for each scenario. Each image scenario inherits the foliation of the biological coordinate system induced by the birth tag τ and enlightened by the color gradient. At each time, a close curve of new coordinates appear (we have $X = [0, 1] \times \mathbb{S}^1$).

and progressively developed in the ambient space, leaf by leaf. In order to ensure that a scenario generated by q globally corresponds to the trivial scenario induced by the coordinate system, we want each leaf of X to be embedded in $\mathbb{R}^d$. Some exceptions are yet allowed, typically for the outer leaves $X_{\{0\}}$ and $X_{\{1\}}$, e.g. to model the tip of the horn. We assume thus that

(H^{q_0})   The restriction of q_0 to an inner leaf B_t ($t \in\ ]0, 1[$) is a smooth immersion between B_t and $\mathbb{R}^d$.

This development *leaf by leaf* induced by the growth dynamic raises some difficulties to control the regularity of the final shape. Let us first recall that the existence and uniqueness of the GMEs generated by time-varying vector fields with the growth dynamic have been established in [23] in a general situation that we recall here. It requires a set of *admissibility* conditions on V

(H_1^V) $\left|\begin{array}{l} i)\ V \subset \mathcal{C}^2(\mathbb{R}^d, \mathbb{R}^d)\,. \\[2mm] ii)\ \text{There exists } c > 0 \text{ such that for any } v \in V \text{ and any } x \in \mathbb{R}^d, \\[2mm] \quad \begin{cases} |v(x)|_{\mathbb{R}^d} \leq c|v|_V(|x|_{\mathbb{R}^d} + 1)\,, \\ |dv(x)|_\infty + |d^2v(x)|_\infty \leq c|v|_V\,. \end{cases} \end{array}\right.$

$$\text{(2.1)}$$

Consider then X a compact smooth manifold with corners and $\tau : X \to [0, 1]$. Due to the spatial discontinuity of the indicator function, the ODE of the growth dynamic

$$\dot{q}_t(x) = \mathbb{1}_{\tau(x) \leq t}\, v_t(q_t(x)) \tag{2.2}$$

is defined in the Banach space $L^\infty(X, \mathbb{R}^d)$ and we have the following theorem

Theorem 2.1: *Under the (H_1^V) conditions, for any initial condition $q_0 \in L^\infty(X, \mathbb{R}^d)$ and any control $v \in L^2([0,1], V)$, there exists a unique solution q that satisfies 2.2 a.e. This solution is absolutely continuous: $q \in \mathcal{AC}([0,1], L^\infty(X, \mathbb{R}^d))$ and for any $t \in [0,1]$,*

$$q_t = q_0 + \int_0^t \mathbb{1}_{s \le \tau} v_s \circ q_s ds\,.$$

Corollary 2.2: *In the setting of Theorem 2.1, if $q_0 \in \mathcal{C}(X, \mathbb{R}^d)$ and $\tau \in \mathcal{C}(X, [0,1])$, then for any $v \in L^2([0,1], V)$, $q \in \mathcal{AC}([0,1], \mathcal{C}(X, \mathbb{R}^d))$.*

The spatial regularity of q_t at higher order requires a stronger temporal regularity of v. Even when the initial condition q_0 is of class $\mathcal{C}^\infty$, if v belongs to L_V^2, we can only show that q_1 is differentiable almost everywhere. Figure 4 illustrates how the discontinuity of v impacts the generated shape (rectifiable yet on this basic example, but not $\mathcal{C}^1$). A strong discontinuity of v at time t, meaning that t is not a Lebesgue point of v, affects the regularity of the shape along the leaf created at time t and this irregularity will hold until the end of the scenario, meaning that the final shape q_1 will be irregular along the leaf $q_1(X_{\{t\}})$. However, if v is time continuous, then q_1 is of class $\mathcal{C}^1$:

Proposition 2.3: *In the setting of Theorem 2.1, if $q_0 \in \mathcal{C}^1(X, \mathbb{R}^d)$, $\tau \in \mathcal{C}^1(X, [0,1])$, and $v \in \mathcal{C}([0,1], V)$, then q_1 is of class $\mathcal{C}^1$ and its differential is given for any $x \in X$ by*

$$dq_1(x) = d\phi_{\tau(x),1}(q_0(x)) \circ \big(dq_0(x) - v_{\tau(x)}(q_0(x))d\tau(x)\big), \qquad (2.3)$$

where $\phi_{s,t}$ is the flow of v on the ambient space $\mathbb{R}^d$.

See Proposition 4.8 in [23] for the proof. Note that it is well-known that under the (H_1^V) conditions, this flow is of class $\mathcal{C}^1$ and has a continuous differential in time and space [15, 50].

The lack of spatial regularity of the final shape $q_1(X)$ prevents the usual straightforward identification with a current or a varifold that yet applies in the restrictive setting of Proposition 2.3. We will therefore extend their definitions based on continuous trajectories $v \in \mathcal{C}([0,1], V)$ (setting of Proposition 2.3) to all the solutions generated by $L^2([0,1], V)$. In the next sections, we will present for the purpose of this paper two specific situations where the shapes can be modeled by varifolds and finally we will prove that

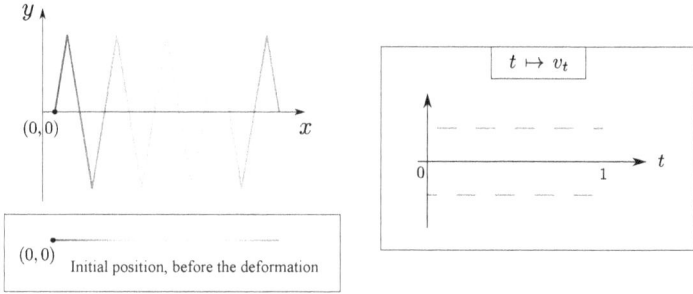

Fig. 4. The final state $q_1(X)$ displayed on the top left is a serrated curve with as many discontinuities as its associated vector field v given on the right as real-valued function modeling vertical translations upwards and downwards. The initial position $q_0(X)$ is a segment.

a similar extension holds for currents in a more general situation. In all these cases of interest, the key is to decompose the shape into its leaves, i.e. with the foliation induced by the birth tag, and to use the density $\mathcal{C}([0,1], V)$ in $L^2([0,1], V)$.

2.2. *Extension for varifold representation*

As presented by Charon and Trouvé [8], k-dimensional smooth shapes embedded in $\mathbb{R}^d$ are modeled by the dual of a reproducing kernel Hilbert space (RKHS) W on $\mathcal{C}_0(\mathbb{R}^d \times G_k(\mathbb{R}^d), \mathbb{R})$ where $G_k(\mathbb{R}^d)$ is the Grassmannian of all k-dimensional subspace of $\mathbb{R}^d$, e.g. lines through the origin of $\mathbb{R}^d$ when $k = 1$. W' represents then a space of varifolds. In all generality, the varifold $\mu \in W'$ associated to a smooth shape Γ is given for any function $\omega \in W$ by

$$\mu(\omega) = \int_{\mathbb{R}^d \times G_k(\mathbb{R}^d)} \omega(x, V) d\mu(x, V) = \int_{\Gamma} \omega(x, T_x\Gamma) d\mathcal{H}^k(x), \qquad (2.4)$$

where for any Borel subset $A \subset \mathbb{R}^d \times G_k(\mathbb{R}^d)$, $\mu(A) = \mathcal{H}^k(\{y \in \Gamma \mid (y, T_y\Gamma) \in A\})$.

The kernel of the RKHS W is denoted k_W and given by the tensor product $k_E \otimes k_T$ of a kernel $k_E(x, y)$ on the ambient space $\mathbb{R}^d$ and a kernel $k_T(U, V)$ on the Grassmannian $G_k(\mathbb{R}^d)$. Throughout this paper, we will consider a degenerate kernel $k_T \equiv 1$ on the Grassmannian and we will write $\omega(x)$ instead of $\omega(x, V)$ and $d\mu(x)$ instead of $d\mu(x, V)$.

2.2.1. *2D case*

We aim for a particularly simple situation to produce a counterexample. The shape is a horn modeled by a curve in $\mathbb{R}^2$. With the previous notation, we would define the coordinate space by $X = [0,1] \times \{-1,1\}$ and the birth tag τ by the projection on the first coordinate. However, to simplify the notation, we project X on the interval $\Omega = [-1,1]$ with $(t,b) \mapsto t \times b$, which merge the two initial coordinates that will model the tip of the horn. The curve is then parameterized by Ω and the initial condition is given by $q_0(r) = (r,0)$. The horn is thus initially flattened on the horizontal segment $[-1,1] \times \{0\} \subset \mathbb{R}^2$. The birth tag is defined for any $r \in \Omega$ by $\tau(r) \doteq |r|$. This means that points are progressively displaced starting from the points at the center and ending with the points at the boundaries of the segment. The deformations are reduced to vertical translations and the space of vector fields V is canonically identified to $\mathbb{R}$. For any $v \in L^2([0,1],\mathbb{R})$, the growth dynamic is given by

$$\dot{q}_t(r) = (0, v(t))\mathbf{1}_{|r| \leq t}$$

and $q_t(r) = (r, \mathbf{1}_{|r| \leq t} \int_{|r|}^{t} v(s)ds)$. The object of interest in this section is the shape at its final age $t = 1$ given by $q_1(r) = \gamma_v(r)$ where $\gamma_v : [-1,1] \to \mathbb{R}^2$ is defined by

$$\gamma_v(r) = \left(r, \int_{|r|}^{1} v(s)ds\right). \tag{2.5}$$

Note that the curve γ_v is symmetric about the vertical axis $\{0\} \times \mathbb{R}$. Examples are displayed hereafter in Figure 6.

For any $v \in \mathcal{C}([0,1],\mathbb{R})$, the varifold associated to the curve γ_v is denoted $\mu_v \in W'$ and it is defined for any $\omega \in W$ by

$$\mu_v(\omega) = \int_{\Gamma_v} \omega(x)d\mathcal{H}^1(x) = \int_{-1}^{1} \omega(\gamma_v(r))|\dot{\gamma}_v(r)|dr$$

$$= \int_{-1}^{1} \omega(\gamma_v(r))\sqrt{1 + v(|r|)^2}\,dr, \tag{2.6}$$

where $\dot{\gamma}_v(r)$ is defined for any $r \in \Omega$ such that $\tau(r) \notin \{0,1\}$, i.e. $r \notin \{-1,0,1\}$ (which corresponds to the boundary of X with the initial notation). We have $\dot{\gamma}_v(r) = (1, -v(r))$ if $r \in]0,1[$ and $\dot{\gamma}_v(r) = (1, v(-r))$ if $r \in]-1,0[$. When $v \in L^2([0,1],\mathbb{R})$, $\dot{\gamma}_v$ is only defined a.e. The application

$$\Psi_W : \left(L^2([0,1],V), |\cdot|_{L^2_V}\right) \longrightarrow \qquad W'$$
$$v \qquad \longmapsto \mu_v : \omega \mapsto \int_{-1}^{1} \omega(\gamma_v(r))\sqrt{1 + v(|r|)^2}\,dr \tag{2.7}$$

is yet well defined and continuous $(W \hookrightarrow \mathcal{C}_0(\mathbb{R}^2 \times G_1(\mathbb{R}^2), \mathbb{R}))$. Moreover, since $\mathcal{C}([0,1], V)$ is dense in $L^2([0,1], V)$, Ψ_W is the unique continuous extension of its restriction to $\mathcal{C}([0,1], V)$.

2.2.2. *3D case*

The coordinate space X is now a cylinder but to simplify the notation, we instead parametrize the shapes by Ω the unit disc, equipped with the polar coordinate system. Points at their initial position are given by $q_0(\theta, r) = (r \cos \theta, r \sin \theta, 0)$. The birth tag τ is equal to the radius $\tau(\theta, r) = r$. The growth dynamic is as before limited to vertical translations:

$$v \in L^2([0,1], \mathbb{R}), \qquad \dot{q}_t(\theta, r) = (0, 0, v(t)) \mathbb{1}_{r \leq t}.$$

The energy only refers to the final state of the shape. Thus, defining $\gamma_v(\theta, r) = q_1(\theta, r)$, it follows that any time-varying vector field $v \in L^2([0,1], \mathbb{R})$ generates a surface described by the parametric function

$$\gamma_v(\theta, r) = \left(r \cos \theta, r \sin \theta, \int_r^1 v_s \, ds \right). \tag{2.8}$$

Let $\mathrm{J}\gamma_v$ be the Jacobian determinant of γ_v,

$$\partial_\theta \gamma_v(\theta, r) = (-r \sin \theta, r \cos \theta, 0), \quad \mathrm{J}\gamma_v(\theta, r) = |\partial_\theta \gamma_v(\theta, r) \wedge \partial_r \gamma_v(\theta, r)|$$
$$\partial_r \gamma_v(\theta, r) = (\cos \theta, \sin \theta, -v(r)), \qquad\qquad = r \sqrt{1 + v(r)^2}.$$

For any $v \in \mathcal{C}([0,1], \mathbb{R})$, the linear form $\mu_v \in W'$ that represents the surface is given for any $\omega \in W$ by

$$\mu_v(\omega) = \int_0^{2\pi} \int_0^1 \omega(\gamma_v(\theta, r)) r \sqrt{1 + v(r)^2} \, dr d\theta.$$

Once again this expression can be extended to $v \in L^2([0,1], V)$ and it defines the unique extension of $v \mapsto \mu_v$.

2.3. *Extension for current representation*

Let Γ be a smooth oriented k-dimensional submanifold and denote for any $x \in \Gamma$, $(T_1(x), \ldots, T_k(x))$ an orthonormal oriented basis of the tangent space $T_x\Gamma$. Then Γ is identified to the current $\mu \in \mathcal{C}_0(\mathbb{R}^d, (\bigwedge^k \mathbb{R}^d)^*)'$ defined for any $\omega \in \mathcal{C}_0(\mathbb{R}^d, (\bigwedge^k \mathbb{R}^d)^*)$ by

$$\mu(\omega) = \int_\Gamma \omega(x)(T_1(x) \wedge \cdots \wedge T_k(x)) d\mathcal{H}^k(x). \tag{2.9}$$

This definition can also be used for rectifiable sets which includes submanifolds with corners. We can now define the current associated to the final mapping q_1 generated by a continuous vector field.

Definition 2.4: For any $v \in \mathcal{C}([0,1], V)$, the current associated to the mapping $q_1 : X \to \mathbb{R}^d$ is defined for any $\omega \in C_0(\mathbb{R}^d, (\bigwedge^k \mathbb{R}^d)^*)$ by

$$\mu_v(\omega) = \int_X q_1^* \omega = \int_X \omega(q_1(x)) \left(\frac{\partial q_1}{\partial x_1}(x) \wedge \cdots \wedge \frac{\partial q_1}{\partial x_k}(x) \right) dx_1 \cdots dx_k . \tag{2.10}$$

For this purpose we rewrite equation 2.10 with the foliation of X given by its tagging function τ. Consider $X = [0,1] \times B$ where B is an oriented compact manifold with corners so that τ is just the projection on the first coordinate (for any $(t, b) \in [0,1] \times B$, $\tau(t, b) = t$). Denote $(Y_t)_{0 < t < 1}$ the set of submanifolds of $\mathbb{R}^d$ that are the images of $B_t \doteq \{t\} \times B$ by q_0. By definition of the growth dynamic, for any $x \in X$ and any $t \geq \tau(x)$,

$$q_t(x) = q_0(x) + \int_0^t \mathbb{1}_{s \leq \tau(x)} v_s(q_s(x)) ds = \phi_{\tau(x),t}(q_0(x)) . \tag{2.11}$$

The image of q_1 can then be rewritten

$$q_1(X) = \bigcup_{t \in [0,1]} \phi_{t,1}(q_0(B_t)) = \bigcup_{t \in [0,1]} \phi_{t,1}(Y_t) . \tag{2.12}$$

We recall that for almost every $t \in [0,1]$ the restriction $q_0 : B_t \to Y_t$ is a $\mathcal{C}^1$ diffeomorphism ((H^{q_0}) condition).

We can now extend for any L_V^2-scenario the definition of the current associated to its final age. The proof lies on the fact that almost all the restrictions of q_1 to the leaves $X_{\{t\}}$ are of class $\mathcal{C}^1$.

Proposition 2.5: *The function $v \mapsto \mu_v$ defined for $v \in \mathcal{C}([0,1], V)$ has a unique continuous extension*

$$\left(L^2([0,1], V), | \cdot |_{L_V^2} \right) \longrightarrow \left(C_0(\mathbb{R}^d, (\Lambda^k \mathbb{R}^d)^*), | \cdot |_\infty \right)^* \tag{2.13}$$
$$v \longmapsto \mu_v : \omega \mapsto \int_0^1 \left[\int_{Y_t} \iota_{(h_t - v_t)} \phi_{t,1}^* \omega \right] dt ,$$

where $(\phi_{s,t})_{s \leq t}$ is the flow of v, $\phi_{t,1}^* \omega$ is the pullback of ω by $\phi_{t,1}$, ι is the interior product and h_t is the unique vector field on Y_t defined for almost any $t \in [0,1]$ and any $x \in B_t$ by $h_t(q_0(x)) = \frac{\partial q_0}{\partial t}(x)$.

Proof: Let us call here φ the application $v \mapsto \mu_v$ given by Definition 2.4 when $v \in \mathcal{C}([0,1], V)$ and $\overline{\varphi}$ the application defined here by equation 2.13.

We will first show that φ and $\overline{\varphi}$ coincides on $v \in \mathcal{C}([0,1], V)$. Then, we will show that $\overline{\varphi}$ is indeed a continuous linear application.

We decompose $X = [0,1] \times B$ with a partition of unity of B. Hence, we just have to consider the case of a support $[0,1] \times U$ where (U, ψ) is a coordinate chart around a point $b \in B$ (consistent with the orientation). We can thus define a local coordinate system $x = (t, b^1, \ldots, b^{k-1})$ on $[0,1] \times U$ and we have $\frac{\partial q_1}{\partial b^i}(x) = d\phi_{t,1}(q_0(x)) \circ \frac{\partial q_0}{\partial b^i}(x)$ and $\frac{\partial q_1}{\partial t}(x) = d\phi_{t,1}(q_0(x)) \circ \left(\frac{\partial q_0}{\partial t}(x) - v_t(q_0(x))\right) = d\phi_{t,1}(q_0(x)) \circ (h_t - v_t)(q_0(x))$. Therefore,

$$\int_{[0,1] \times U} q_1^* \omega = \int_{[0,1] \times U} \omega(q_1(t,b)) \left(\frac{\partial q_1}{\partial t}(t,b) \bigwedge_{i=1}^{k-1} \frac{\partial q_1}{\partial b^i}(t,b)\right) db^1 \cdots db^{k-1} dt$$

$$= \int_{[0,1] \times U} \omega(\phi_{t,1}(q_0(t,b)))$$

$$\left(d\phi_{t,1}(q_0(t,b)) \circ (h_t - v_t)(q_0(t,b)) \bigwedge_{i=1}^{k-1} d\phi_{t,1}(q_0(t,b)) \circ \frac{\partial q_0}{\partial b^i}(t,b)\right) db^1 \cdots db^{k-1} dt$$

$$= \int_0^1 \left[\int_B (\phi_{t,1}^* \omega)(q_0(t,b)) \left((h_t - v_t)(q_0(t,b)) \bigwedge_{i=1}^{k-1} \frac{\partial q_0}{\partial b^i}(t,b)\right) db^1 \cdots db^{k-1}\right] dt$$

$$= \int_0^1 \left[\int_{Y_t} \iota_{(h_t - v_t)} \phi_{t,1}^* \omega\right] dt.$$

Now, we have $\sup_{t \in [0,1]} |d\phi_{t,1}| = g_1(|v|_{L_V^2}^2)$, dq_0 is also bounded on X, so that h_t and v_t are bounded on $q_0(X)$ and therefore

$$\left|\int_0^1 \left[\int_{Y_t} \iota_{(h_t - v_t)} \phi_{t,1}^* \omega\right] dt\right| \leq |\omega|_\infty g_2(|v|_{L_V^2}^2), \tag{2.14}$$

where g_1 and g_2 are increasing functions independent of v and ω. Consequently, for any $v \in L_V^2$, $\overline{\varphi}(v) = \mu_v$ belongs to $\mathcal{C}_0(\mathbb{R}^d, (\Lambda^k \mathbb{R}^d)^*)^*$ and $\overline{\varphi}$ is continuous due to the regularity of the flow and of the interior product. Hence, since $\mathcal{C}([0,1], V)$ is dense in L_V^2, $\overline{\varphi}$ is the unique continuous extension of φ. $\qquad\square$

Remark 2.6: Note that μ_v is not exactly the current associated to the image $q_1(X)$. Indeed, even if q_1 is differentiable, it might not be an embedding. Two counter-examples are presented in Figure 5. In the first case, the direction of the development is suddenly reversed twice so that the curve is folding on itself. Hence, if we refer to number of preimages of each point of $q_1(X)$ as a thickness of the shape, then the thickness here is equal to 1

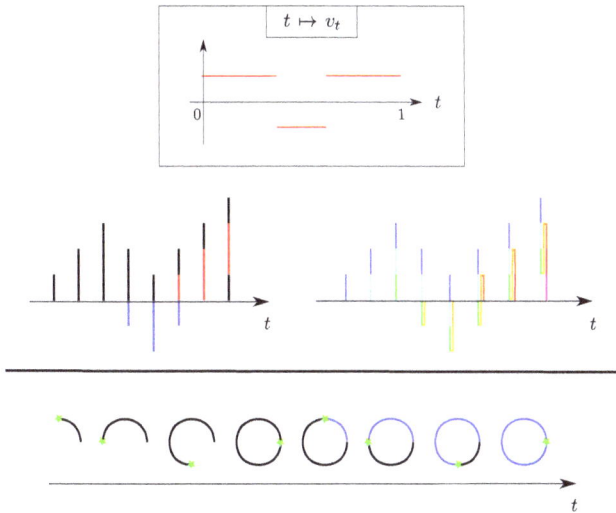

Fig. 5. Two examples of scenarios. In both situations, X is a segment but $q_0(X)$ is reduced to a point. On top, v is given by piecewise constant vertical translations upwards and downwards, modeled by real-valued function. The final image is a segment but when v changed its sign, the curve folded on itself. The scenario is displayed again on the left but we slightly separated the multiple fibers of the curve. One can think to a magic trick where colored attached strings are pulling out from the initial position point. On the bottom, the scenario is generated by a constant rotation anticlockwise. The ambient space is exactly rotated twice during the time interval $[0, 1]$. We display the development of the curves with three colors depending of the thickness: dark for 1, blue for 2 and red for 3. The green star on the bottom is just displayed to highlight the evolution of one specific point.

or 3. On the second case, the curve completely overwrites itself, so that the thickness is equal to 2 on each point.

This phenomenon depends on q_0 and v and cannot be anticipated. The current associated to our shapes therefore counts these repetitions. However, in the first scenario, since the orientation is reversed twice and by linearity of the currents with respect to the tangential data, the repetition is canceled and we have $\mu_v(\omega) = \int_{q_1(X)} \omega$. On the second example, the orientation is the same on each layer so that $\mu_v(\omega) = 2\int_{q_1(X)} \omega$. At last, note that in practice, these situations should not happen with optimal vector fields. The penalization of v should prevent these artifacts. Generating a cancel effect via an overlapping should induce an additional cost on v with yet no reduction of the data attachment term since the current would be the same without this overlapping. Likewise, the gain of thickness as in the second example is necessarily taken from spatial correspondences with the target shape and should therefore not be profitable (at least for a metric with a reasonable scale so that the position of the points are enough discriminated).

3. Discontinuity for varifold data term

This section highlights a counterexample to the existence of continuous minimizers of the energy when the attachment term is build on a space of varifolds. We present a first counterexample in a 2D case that will then be adapted to a 3D case.

3.1. *Counterexample with curves*

The growth model is here given by the setting of Section 2.2.1 where we recall that the final shape $\gamma_v : [-1, 1] \to \mathbb{R}^2$ of a growth scenario is defined by

$$\gamma_v(r) = (r, \int_{|r|}^{1} v(s)ds) \tag{3.1}$$

and it is represented by a varifold denoted μ_v. We also consider a target horn $\gamma_{v^{\text{tar}}} : [-1, 1] \to \mathbb{R}^2$ given by the same equation for a time-varying vector field $v^{\text{tar}} \in L^2([0, 1], \mathbb{R})$. The discrepancy between the two curves γ_v and $\gamma_{v^{\text{tar}}}$ is then estimated by the distance between μ_v and $\mu_{v^{\text{tar}}}$ in W', this is to say with the norm $|\mu_v - \mu_{v^{\text{tar}}}|_{W'}$.

Finally, the matching problem consists in minimizing the energy given by the sum of a penalization term on v and this data attachment term

$$E_W^\lambda(v) \doteq \frac{1}{2} \int_0^1 v(r)^2 dr + \frac{\lambda}{2}|\mu_v - \mu_{v^{\text{tar}}}|_{W'}^2 . \tag{3.2}$$

Our aim is to examine whether the regularization L^2 on v and the data attachment term on varifolds ensure the continuity of global minimizers of E_W^λ. We will prove the following theorem:

Theorem 3.1: *There exist $v^{\text{tar}} \in L^2([0, 1], \mathbb{R})$, $\lambda > 0$ and W such that no global minizer v^* of E_W^λ given by 3.2 is a continuous function on $[0, 1]$. Moreover, one can assume that $v^{\text{tar}} \in \mathcal{C}^\infty([0, 1], \mathbb{R})$.*

We will consider a perturbation parameterized by $\epsilon \geq 0$ of a degenerate constant kernel $k_W \equiv 1$. The solutions of the optimization problem associated to this kernel will be especially easy to explicit.

Definition 3.2: Define $k_T \equiv 1$ and a set of kernels $k_\epsilon(x,y) = \rho(\epsilon|x-y|^2_{\mathbb{R}^2})$ where ρ is a positive function such that $\rho(0) = 1$, $\dot{\rho}$ is bounded on $\mathbb{R}$ and $\dot{\rho}(0) < 0$. They generate a set of kernels $k_\epsilon \otimes k_T$ on $\mathbb{R}^2 \times G_1(\mathbb{R}^2)$ that do not see the tangential directions. Each kernel $k_\epsilon \otimes k_T$ for $\epsilon \geq 0$ produces a RKHS denoted W_ϵ. Since W_ϵ depends on ϵ, the energy will be denoted $E^\lambda(\epsilon, v)$ to refer to $E^\lambda_{W_\epsilon}(v)$.

This construction could probably be extended to a symmetric situation with a perturbation $k'_{\epsilon'}$ of k_T. Note yet that it would require to investigate the spatial regularity of the curve γ_v. Hence, we will only consider $k_T \equiv 1$.

3.1.1. *Solutions for the degenerate kernel*

The first step is to study the minimizers of $E^\lambda(0, \cdot)$. When $\epsilon = 0$, the kernel $k_{W_0} = k_0 \otimes k_T$ is constant and W_0 is a 1-dimensional space whose elements w are all constant. In this case, the expression of the data attachment term is particularly simple:

$$|\mu_v - \mu_{v^{\text{tar}}}|^2_{W'_0} = \iint_{\mathbb{R}^2 \times \mathbb{R}^2} k_0(x,y) d(\mu_v - \mu_{v^{\text{tar}}})(x) d(\mu_v - \mu_{v^{\text{tar}}})(y)$$

$$= \left(\int_{\mathbb{R}^2} 1 d(\mu_v - \mu_{v^{\text{tar}}})(x) \right)^2$$

$$= (\ell(v) - \ell(v^{\text{tar}}))^2 \,,$$

where $\ell(v)$ measures the length of the curve generated by v

$$\ell(v) = 2 \int_0^1 \sqrt{1 + v(t)^2} dt \,. \tag{3.3}$$

Finally, the energy in this case is given by

$$E^\lambda(0, v) = \frac{1}{2} \int_0^1 v(t)^2 dt + \frac{\lambda}{2} (\ell(v) - \ell(v^{\text{tar}}))^2 \,. \tag{3.4}$$

The global minimizers have then an explicit expression given by the following proposition.

Proposition 3.3: *Assume that* $\ell_0 = \frac{4\lambda\ell(v^{\text{tar}})}{4\lambda+1} > 2$. *Then* $v^* \in L^2([0,1], \mathbb{R})$ *is a global minimizer of* $E^\lambda(0, \cdot)$ *if and only if we have at almost all time* $v^*(t)^2 = \ell_0^2/4 - 1$. *In particular, if* $v^* \in \mathcal{C}([0,1], \mathbb{R})$ *then* v^* *is constant.*

Proof: We have the following elementary lemma:

Lemma 3.4: *If* $\ell_0 = \frac{4\lambda\ell(v^{\text{tar}})}{4\lambda+1} > 2$, *then* ℓ_0 *minimizes*

$$P(\ell) \doteq \frac{\ell^2/4 - 1}{2} + \frac{\lambda}{2}(\ell(v^{\text{tar}}) - \ell)^2$$

on $\mathbb{R}$. *Moreover, if* $v \in L^2([0,1], \mathbb{R})$ *satisfies* $v(t)^2 = \ell^2/4 - 1$ *a.e. where* $\ell \in \mathbb{R}$, *then* $E^\lambda(0, v) = P(\ell)$.

Define for $\ell \geq 2$, the function $\rho_\ell : \mathbb{R} \to \mathbb{R}$ by

$$\rho_\ell(z) \doteq \frac{z^2}{2} - \frac{\ell}{2}\sqrt{z^2 + 1}.$$

This function is even, tends to $+\infty$ when $|z|$ tends to $+\infty$ and $\dot{\rho}_\ell(z) = 0 \Leftrightarrow z - \frac{\ell}{2}\frac{z}{\sqrt{z^2+1}} = 0 \Leftrightarrow (z = 0$ or $z^2 = \frac{\ell^2}{4} - 1)$. Therefore, ρ_ℓ admits two minimizers that satisfy

$$\begin{cases} z^2 = \ell^2/4 - 1 > 0 \\ \rho_\ell(z) = (\ell^2/4 - 1)/2 - (\ell/2)\sqrt{\ell^2/4 - 1 + 1} = -(\ell^2/4 + 1)/2. \end{cases}$$

It results that the minimum of

$$R_\ell(v) \doteq \int_0^1 \rho_\ell(v(t))dt$$

is reached at $v^* \in L^2([0,1], \mathbb{R})$ if and only if

$$v^*(t)^2 = \ell^2/4 - 1 \quad \text{a.e..} \tag{3.5}$$

By construction, these minimizers are exactly the solutions of the constrained optimization problem

$$\left| \begin{array}{l} \min_{L^2} \int v(t)^2 dt \\ \text{with } \ell(v) = \ell \end{array} \right. .$$

Indeed, if v^* satisfies equation 3.5, then $\ell(v^*) = 2\int_0^1 \sqrt{v^*(t)^2 + 1}dt = \ell$ and if there exists another $v \in L^2([0,1], \mathbb{R})$ such that $\ell(v) = \ell$ and $\int_0^1 v(t)^2 dt < \int_0^1 v^*(t)^2 dt$ then $R_\ell(v) < R_\ell(v^*)$ which is absurd.

Consequently, any minimizer $v^* \in L^2([0,1], \mathbb{R})$ of $E^\lambda(0, \cdot)$ satisfies $v^*(t)^2 = \ell_0^2/4 - 1$ a.e. where $\ell_0 = \ell(v^*)$ is defined on $[2, +\infty[$ and must minimize $\ell \mapsto E^\lambda(0, v^*) = \frac{\ell^2/4-1}{2} + \frac{\lambda}{2}(\ell(v^{\text{tar}}) - \ell)^2$, i.e. $\ell_0 = \frac{4\lambda\ell(v^{\text{tar}})}{1+4\lambda} > 2$. Moreover, there exist exactly two continuous minimizers in $L^2([0,1], \mathbb{R}) \cap \mathcal{C}([0,1], \mathbb{R})$ given by $v^+ \equiv \sqrt{\ell_0^2/4 - 1}$ and $v^- = -v^+$. $\qquad \square$

Remark 3.5: Note that with the degenerate kernel $k_W \equiv 1$, the energy has continuous global minimizers. However, they are only two of an infinite

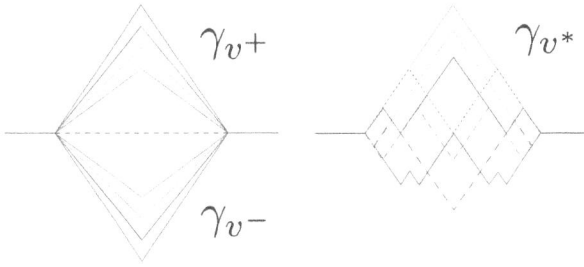

Fig. 6. On the left. Solutions generated by the continuous minimizers v^+ and v^-. Each color is associated to a length ℓ_0. The dotted line is the image of the initial position, i.e. the base of the horn. On the right. Solutions generated by a set of discontinuous minimizers v^* at fixed ℓ_0 (and v^+ on the top).

number of solutions. Figure 6 illustrates on its left the two curves generated by v^+ and v^- for four given lengths ℓ_0. Figure 6 illustrates on its right few examples where ℓ_0 is fixed. The condition to be a minimizer leads to a large set of different type of curves as we only control the length of the final curve. Assume now that the target is some kind of sinusoidal curve. One can then easily see that from a spatial point of view, the two curves γ_{v^+} and γ_{v^-} are probably the less optimal solutions among the complete set of solutions γ_{v^*}. Hence, as soon as the kernel k_W is perturbed and allowed to capture some spatial position of the target, one can expect that the new energies associated to v^+ and v^- are higher than the energy of at least one other solution v^*.

Hypothesis: There exists v^*, such that for any $\epsilon > 0$ small enough,

$$E^\lambda(\epsilon, v^+) > E^\lambda(\epsilon, v^*) \quad \text{and} \quad E^\lambda(\epsilon, v^-) > E^\lambda(\epsilon, v^*).$$

3.1.2. *Perturbation of the degenerate kernel*

The next step to prove the theorem is to investigate the minimizers of $v \mapsto E^\lambda(\epsilon, v)$ where $\epsilon > 0$ and

$$E^\lambda(\epsilon, v) = \frac{1}{2} \int_0^1 v(r)^2 dr + \frac{\lambda}{2} |\mu_v - \mu_{v^{\text{tar}}}|_{W'}^2. \tag{3.6}$$

The following proposition will establish that if some of these minimizers are continuous, they necessarily lie in a neighborhood of v^+ or v^- (the two continuous global minimizers of $E^\lambda(0, v)$). Analyzing the variations of $\epsilon \to E^\lambda(\epsilon, v)$ will then indicate that in some situations these minimizers cannot be global minimizers.

Remark 3.6 (RKHS properties). Denote $\omega = K_W(\mu_v - \mu_{v^{\text{tar}}})$, where $K_W : W' \to W$ is the canonical isomorphism of Hilbert spaces. By construction of a RKHS, ω is given at any $(x, V) \in \mathbb{R}^2 \times G_1(\mathbb{R}^2)$ by

$$\omega(x, V) = \int_{\mathbb{R}^2 \times G_1(\mathbb{R}^2)} k_W\big((x, V), (y, V')\big) d(\mu_v - \mu_{v^{\text{tar}}})(y, V'). \qquad (3.7)$$

Let us recall then that $|\mu_v - \mu_{v^{\text{tar}}}|^2_{W'} = (\mu_v - \mu_{v^{\text{tar}}})(\omega)$ and

$$\frac{\partial}{\partial v} \frac{1}{2} |\mu_v - \mu_{v^{\text{tar}}}|^2_{W'} = \left(\frac{\partial}{\partial v}\mu_v\right)(\omega). \qquad (3.8)$$

Proposition 3.7: *Assume that $\ell_0 = \frac{4\lambda \ell(v^{\text{tar}})}{4\lambda + 1} > 2$. If for any $\epsilon > 0$ small enough, there exists a continuous global minimizer v_ϵ of $E^\lambda(\epsilon, \cdot)$, then*

$$\lim_{\epsilon \to 0} \big(\min(|v_\epsilon - v^+|_\infty, |v_\epsilon - v^-|_\infty)\big) = 0, \qquad (3.9)$$

where $v^+ \equiv \sqrt{\ell_0^2/4 - 1}$ and $v^- = -v^+$ are the only continuous global minimizers of $E^\lambda(0, \cdot)$.

Proof: Denote $\omega_v(\epsilon, \cdot) = K_{W_\epsilon}(\mu_v - \mu_{v^{\text{tar}}})$. Since the kernel k_{W_ϵ} of W_ϵ is reduced to $k_\epsilon(x, y) = \rho(\epsilon |x - y|^2)$, the tangential component of the varifold $\mu_v - \mu_{v^{\text{tar}}}$ can be ignored and we have (see Remark 3.6)

$$\omega_v(\epsilon, \cdot) = \int_{\mathbb{R}^2} k_\epsilon(\cdot, y) d(\mu_v - \mu_{v^{\text{tar}}})(y).$$

With the symmetry of the curve γ_v, the associated varifold μ_v can be rewritten

$$\mu_v(\omega) = \sum_{i=0}^{1} \int_0^1 \omega(S^i(\gamma_v(r))) \sqrt{1 + v(r)^2} dr, \qquad (3.10)$$

where $S^1 = S$ is the symmetry with respect to the vertical axis through origin and $S^0 = \text{Id}$. We then symmetrize ω_v as follows

$$\omega_v^S(\epsilon, x) \doteq \sum_{i=0}^{1} \omega_v(\epsilon, S^i(x)) = \sum_{i=0}^{1} \int_{\mathbb{R}^2} \rho(\epsilon |S^i(x) - y|^2) d(\mu_v - \mu_{v^{\text{tar}}})(y),$$

so that equation 3.10 reads

$$\mu_v(K_{W_\epsilon}(\mu_v - \mu_{v^{\text{tar}}})) = \mu_v(\omega_v(\epsilon, \cdot)) = \int_0^1 \omega_v^S(\epsilon, \gamma_v(r)) \sqrt{v(r)^2 + 1} dr.$$

We recall that $\gamma_v(r) = (r, \int_{|r|}^1 v(s)ds)$ and that for any $v \in L^2$

$$\lim_{\epsilon \to 0} \omega_v^S(\epsilon, \cdot) \equiv 2(\ell(v) - \ell(v^{\text{tar}})). \qquad (3.11)$$

One can easily prove that $v \mapsto E^\lambda(\epsilon, v)$ is differentiable with respect to v and we have for any $\delta v \in L^2$

$$\left(\frac{\partial E^\lambda}{\partial v}(\epsilon, v) \mid \delta v \right) = \int_0^1 v(t)\delta v(t)dt + \lambda \left(\frac{\partial}{\partial v}\mu_v \mid \delta v \right)(K_{W_\epsilon}(\mu_v - \mu_{v^{tar}}))$$

$$\left(\frac{\partial E^\lambda}{\partial v}(\epsilon, v) \mid \delta v \right) = \int_0^1 v(t)\delta v(t)dt$$

$$+ \lambda \int_0^1 \left(\left(\partial_2 w_v^S(\epsilon, \gamma_v(t)) \mid \left(0, \int_t^1 \delta v(s)ds \right) \right) \sqrt{v(t)^2 + 1} + w_v^S(\epsilon, \gamma_v(t)) \frac{v(t)\delta v}{\sqrt{v(t)^2 + 1}} \right) dt \,,$$

where $\partial_2 w_v^S(\epsilon, x)$ is the derivative of w_v^S with respect to x. Denote

$$\alpha_{\epsilon,v}(s) \doteq \int_0^s \left(\partial_2 w_v^S(\epsilon, \gamma_v(t)) \mid (0, 1) \right) \sqrt{v(t)^2 + 1}dt \,,$$

so that

$$\frac{\partial E^\lambda}{\partial v}(\epsilon, v) = \left(1 + \lambda \frac{w_v^S(\epsilon, \gamma_v)}{\sqrt{v^2 + 1}} \right) v + \alpha_{\epsilon,v} \,.$$

Note then that $|\alpha_{\epsilon,v}|_\infty = O(\epsilon)$. Indeed,

$$\partial_2 w_v^S(\epsilon, x) = \epsilon \sum_{i=0}^1 S^i \left(\int_0^1 2(S^i(x) - y)\dot{\rho}(\epsilon|S^i(x) - y|^2)d(\mu_v - \mu_{v^{tar}})(y) \right)$$

and since $\dot{\rho}$ is bounded on $\mathbb{R}$ we deduce that for any bounded neighborhood of $(0,0)$ in $\mathbb{R}^+ \times L^2([0, 1], \mathbb{R})$, we have $|\partial_2 w_v^S(\epsilon, \gamma_v)|_\infty = O(\epsilon)$ and

$$|\alpha_{\epsilon,v}|_\infty = O(\epsilon) \,. \tag{3.12}$$

Assume now that for any $\epsilon > 0$, there exists a continuous solution v_ϵ that minimizes $E^\lambda(\epsilon, \cdot)$. It must then satisfy

$$\left(1 + \lambda \frac{w_{v_\epsilon}(\epsilon, \gamma_{v_\epsilon})}{\sqrt{v_\epsilon^2 + 1}} \right) v_\epsilon + \lambda \alpha_{\epsilon,v_\epsilon} = 0 \text{ a.e.. } \tag{3.13}$$

For ϵ small enough, equations 3.13 and 3.12 imply that there exist $M > 0$ and $\beta_\epsilon \geq 0$ such that at almost any time $t \in [0, 1]$ we have either

$$|(\sqrt{v_\epsilon(t)^2 + 1} - \beta_\epsilon| \leq M\epsilon^{1/2}\sqrt{v_\epsilon(t)^2 + 1} \quad \text{or} \quad |v_\epsilon(t)| \leq M\epsilon^{1/2} \,. \tag{3.14}$$

To go further let us first show that the lengths of the curves $(\gamma_{v_\epsilon})_{\epsilon \geq 0}$ converge.

Lemma 3.8: $\ell(v_\epsilon)$ *tends to* $\ell_0 = \ell(v^+) = \ell(v^-)$.

Proof: We have

$$E^\lambda(0, v_\epsilon) \le E^\lambda(\epsilon, v_\epsilon) + o(1) \le E^\lambda(\epsilon, v^+) + o(1) \le E^\lambda(0, v^+) + o(1).$$
$$(3.15)$$

Left and right inequalities result from the continuity of $E^\lambda(\cdot, v)$. Since v_ϵ minimizes $E^\lambda(\epsilon, \cdot)$, the central inequality is also true. Consider now $\ell_0 = \frac{4\lambda\ell(v^{\text{tar}})}{4\lambda+1} > 2$ and the polynomial $P(\ell) = \frac{1}{2}(\ell^2/4 - 1) + \frac{\lambda}{2}(\ell(v^{\text{tar}}) - \ell)^2$. Lemma 3.4 says that $\ell(v^+) = \ell_0$ and $E^\lambda(0, v^+) = P(\ell(v^+)) = P(\ell_0)$. Moreover, if for any $\epsilon > 0$, we define $\delta v_\epsilon \equiv \sqrt{\ell(v_\epsilon)^2/4 - 1}$, then $\ell(\delta v_\epsilon) = \ell(v_\epsilon)$ and $E^\lambda(0, \delta v_\epsilon) \le E^\lambda(0, v_\epsilon)$ (δv_ϵ minimizes R_ℓ, see proof of Proposition 3.3). It results from equation 3.15 and Lemma 3.4 that

$$P(\ell(v_\epsilon)) = E^\lambda(0, \delta v_\epsilon) \le P(\ell_0) + o(1).$$

At last, since ℓ_0 minimizes P, we have

$$P(\ell_0) \le P(\ell(v_\epsilon)) \le P(\ell_0) + o(1).$$

Hence, since P admits a unique minimizer (quadratic polynomial), we have $\ell(v_\epsilon) = \ell_0 + o(1) = \ell(v^+) + o(1)$. $\square$

We can now prove that the first case of equation 3.14 is the only one true. Denote for any $\epsilon > 0$, $A_\epsilon \doteq \{t \in [0,1] \,|\, |\sqrt{v_\epsilon(t)^2 + 1} - \beta_\epsilon| \le O(\epsilon^{1/2})\sqrt{v_\epsilon(t)^2 + 1}\}$ and $\ell_\epsilon \doteq 2(\lambda_\mathbb{R}(A_\epsilon)\beta_\epsilon + (1 - \lambda_\mathbb{R}(A_\epsilon)))$ (where $\lambda_\mathbb{R}$ is the Lebesgue measure). Then

$$\ell(v_\epsilon) = 2\int_0^1 \sqrt{v_\epsilon(t)^2 + 1}\,dt = 2\left(\int_{A_\epsilon}\beta_\epsilon dt + \int_{[0,1]\setminus A_\epsilon} 1\,dt + O(\epsilon^{1/2})\right) = \ell_\epsilon + O(\epsilon^{1/2}).$$

Lemma 3.8 implies then that $\ell_\epsilon = \ell_0 + o(1)$. Moreover, according to equation 3.11, we have necessarily $\beta_\epsilon = 2\lambda(\ell(v^{\text{tar}}) - \ell(v_\epsilon)) + o(1)$ so that $\beta_\epsilon = 2\lambda(\ell(v^{\text{tar}}) - \ell_\epsilon) + o(1) = 2\lambda(\ell(v^{\text{tar}}) - \ell_0) + o(1)$. At last, from Proposition 3.3, we have $\ell(v^{\text{tar}}) - \ell_0 = \ell_0/(4\lambda)$ so that $2\beta_\epsilon = \ell_0 + o(1) = \ell_\epsilon + o(1)$. Finally, we deduce by definition of ℓ_ϵ that $\lambda_\mathbb{R}(A_\epsilon) = 1 + o(1)$.

Therefore, there exists $M' > 0$ such that for almost any $t \in [0,1]$,

$$|v_\epsilon(t)^2 - (\ell_0^2/4 - 1)| \le M'\epsilon.$$

Since v_ϵ is continuous and $\ell_0 > 2$, it follows that v_ϵ satisfies either

$$|v_\epsilon - v^+|_\infty \le M'\epsilon \quad \text{or} \quad |v_\epsilon - v^-|_\infty \le M'\epsilon.$$

And finally,

$$\lim_{\epsilon \to 0}\left(\min(|v_\epsilon - v^+|_\infty, |v_\epsilon - v^-|_\infty)\right) = 0.$$
                                                                              $\square$

The final step to prove the theorem is to study the variations of $E^\lambda(\cdot, v)$ with respect to ϵ at a global minimizer $v = v^*$. The aim is to show that the energy around v^+ and v^- increases too fast, with respect to ϵ, to allow any v in their neighborhood to be a global minimizer of $E^\lambda(\epsilon, \cdot)$. As announced in Remark 3.5, the idea is to compare the geometric properties of all the minimizers of $E^\lambda(0, \cdot)$. We will thus rewrite the gradient of this energy via some geometric descriptors.

Definition 3.9: Denote x_v the *centroid* of the curve γ_v defined by

$$x_v \doteq \frac{1}{\ell(v)} \int_{\mathbb{R}^2} x d\mu_v(x) \tag{3.16}$$

and $V(v)$ the associated *variance* defined by

$$V(v) \doteq \frac{1}{\ell(v)} \int_{\mathbb{R}^2} |x - x_v|^2 d\mu_v(x). \tag{3.17}$$

Lemma 3.10: *The function $\epsilon \to E^\lambda(\epsilon, v)$ is differentiable and for $\epsilon = 0$, we have*

$$\frac{\partial E^\lambda}{\partial \epsilon}(0, v) =$$

$$-\lambda\dot\rho(0)\bigg(\Big(\ell(v^{\mathrm{tar}}) - \ell(v)\Big)\Big(\ell(v)V(v) - \ell(v^{\mathrm{tar}})V(v^{\mathrm{tar}})\Big) + \ell(v)\ell(v^{\mathrm{tar}})|x_{v^{\mathrm{tar}}} - x_v|^2\bigg).$$

Proof: The proof depends neither on the dimension of the ambient space nor on the dimension of the varifolds. Let assume that the ambient space is $\mathbb{R}^d$ and let us start to establish with varifolds the algebraic formulae for the variance ($V(X) = E[X^2] - E[X]^2$). For any $v \in L^2$, we have

$$\ell(v)V(v) = \int_{\mathbb{R}^d} |x - x_v|^2 d\mu_v(x)$$

$$= \int_{\mathbb{R}^d} |x|^2 + |x_v|^2 - 2\langle x, x_v\rangle \, d\mu_v(x)$$

$$= \int_{\mathbb{R}^d} |x|^2 \, d\mu_v(x) + \ell(v)|x_v|^2 - 2\langle x_v, \int_{\mathbb{R}^d} x \, d\mu_v(x)\rangle$$

$$= \int_{\mathbb{R}^d} |x|^2 \, d\mu_v(x) - \ell(v)|x_v|^2.$$

Then, one can easily show that $\epsilon \mapsto E^\lambda(\epsilon, v)$ is differentiable and that

$$
\frac{\partial E^\lambda}{\partial \epsilon}(0, v) = \frac{\partial}{\partial \epsilon} \frac{\lambda}{2} |\mu_v - \mu_{v^{\text{tar}}}|_{W_\epsilon'}^2 \Big|_{\epsilon=0}
$$

$$
= \frac{\partial}{\partial \epsilon} \frac{\lambda}{2} \iint_{\mathbb{R}^d \times \mathbb{R}^d} \rho(\epsilon |x - y|^2) d(\mu_v - \mu_{v^{\text{tar}}})(x) d(\mu_v - \mu_{v^{\text{tar}}})(y) \Big|_{\epsilon=0}
$$

$$
= \frac{\lambda}{2} \dot\rho(0) \iint_{\mathbb{R}^d \times \mathbb{R}^d} |x - y|^2 d(\mu_v - \mu_{v^{\text{tar}}})(x) d(\mu_v - \mu_{v^{\text{tar}}})(y)
$$

$$
= \frac{\lambda}{2} \dot\rho(0) \iint_{\mathbb{R}^d \times \mathbb{R}^d} (|x|^2 + |y|^2 - 2\langle x, y\rangle) d(\mu_v - \mu_{v^{\text{tar}}})(x) d(\mu_v - \mu_{v^{\text{tar}}})(y)
$$

$$
= \lambda \dot\rho(0) \bigg(\underbrace{\big(\ell(v) - \ell(v^{\text{tar}})\big) \int_{\mathbb{R}^d} |x|^2 d(\mu_v - \mu_{v^{\text{tar}}})(x)}_{a} - \underbrace{\bigg| \int_{\mathbb{R}^d} x d(\mu_v - \mu_{v^{\text{tar}}})(x) \bigg|^2}_{b} \bigg).
$$

The terms denoted by a and b can be rewritten as follows:

$$
a = \big(\ell(v) - \ell(v^{\text{tar}})\big) \int_{\mathbb{R}^d} |x|^2 d(\mu_v - \mu_{v^{\text{tar}}})(x)
$$

$$
= \big(\ell(v) - \ell(v^{\text{tar}})\big) \bigg(\int_{\mathbb{R}^d} |x|^2 d\mu_v(x) - \int_{\mathbb{R}^d} |x|^2 d\mu_{v^{\text{tar}}}(x) \bigg)
$$

$$
= \big(\ell(v) - \ell(v^{\text{tar}})\big) \big(\ell(v) V(v) - \ell(v^{\text{tar}}) V(v^{\text{tar}})\big)
$$
$$
+ \ell(v)^2 |x_v|^2 - \ell(v)\ell(v^{\text{tar}}) |x_v|^2 + \ell(v^{\text{tar}})^2 |x_{v^{\text{tar}}}|^2 - \ell(v)\ell(v^{\text{tar}}) |x_{v^{\text{tar}}}|^2
$$

and

$$
b = \bigg| \int_{\mathbb{R}^d} x d(\mu_v - \mu_{v^{\text{tar}}})(x) \bigg|^2
$$

$$
= \bigg| \int_{\mathbb{R}^d} x d\mu_v(x) - \int_{\mathbb{R}^d} x d\mu_{v^{\text{tar}}}(x) \bigg|^2
$$

$$
= \big| \ell(v) x_v - \ell(v^{\text{tar}}) x_{v^{\text{tar}}} \big|^2
$$

$$
= \ell(v)^2 |x_v|^2 + \ell(v^{\text{tar}})^2 |x_{v^{\text{tar}}}|^2 - 2\ell(v)\ell(v^{\text{tar}}) \langle x_v, x_{v^{\text{tar}}}\rangle .
$$

Then $a - b$ is equal to

$$
a - b = \big(\ell(v) - \ell(v^{\text{tar}})\big) \big(\ell(v) V(v) - \ell(v^{\text{tar}}) V(v^{\text{tar}})\big)
$$
$$
- \ell(v)\ell(v^{\text{tar}}) \big(|x_v|^2 + |x_{v^{\text{tar}}}|^2 - 2\langle x_v, x_{v^{\text{tar}}}\rangle \big)
$$
$$
= - \big(\ell(v^{\text{tar}}) - \ell(v)\big) \big(\ell(v) V(v) - \ell(v^{\text{tar}}) V(v^{\text{tar}})\big)
$$
$$
- \ell(v)\ell(v^{\text{tar}}) \big| x_v - x_{v^{\text{tar}}} \big|^2 .
$$

We retrieve the announced formula. $\square$

We now exhibit a condition to the existence of a sequence $(v_{\epsilon_n})_n \subset L^2$ such that $\epsilon_n \to 0$ and for any $n \geq 0$, v_{ϵ_n} is a continuous global minimizer of $E^\lambda(\epsilon_n, .)$.

Proposition 3.11: *Assume that $\ell_0 = \frac{4\lambda\ell(v^{\text{tar}})}{4\lambda+1} > 2$. If there exists a decreasing sequence $\epsilon_n \to 0$ such that v_{ϵ_n} is a continuous global minimizer of $E^\lambda(\epsilon_n, 0)$ then for any global minimizer v^* of $E^\lambda(0, \cdot)$, we have*

$$\min\left(\frac{\partial E^\lambda}{\partial \epsilon}(0, v^+), \frac{\partial E^\lambda}{\partial \epsilon}(0, v^-)\right) \leq \frac{\partial E^\lambda}{\partial \epsilon}(0, v^*), \qquad (3.18)$$

where v^+ and v^- are the only two continuous global minimizers of $E^\lambda(0, \cdot)$ (they are constant and defined by $v^+ \equiv \sqrt{\ell_0^2/4 - 1}$ and $v^- = -v^+$).

Proof: Denote $v_n = v_{\epsilon_n}$. According to Proposition 3.7, either v^+ or v^- is an accumulation point of $(v_n)_n$. Assume that $(v_n)_n$ converges to v^+ (one can extract a subsequence if necessary) and consider v^* a global minimizer of $E^\lambda(0, \cdot)$. The continuity of $(\epsilon, v) \mapsto \partial_\epsilon E^\lambda(\epsilon, v)$ on a neighborhood of $(0, v^+)$ implies then that

$$E^\lambda(\epsilon_n, v^*) \geq E^\lambda(\epsilon_n, v_n) = E^\lambda(0, v_n) + \epsilon_n \partial_\epsilon E^\lambda(0, v_n) + o(\epsilon_n)$$
$$= E^\lambda(0, v_n) + \epsilon_n \partial_\epsilon E^\lambda(0, v^+) + o(\epsilon_n)$$
$$\geq E^\lambda(0, v^*) + \epsilon_n \partial_\epsilon E^\lambda(0, v^+) + o(\epsilon_n)$$
$$\geq E^\lambda(\epsilon_n, v^*) - \epsilon_n \partial_\epsilon E^\lambda(0, v^*) + \epsilon_n \partial_\epsilon E^\lambda(0, v^+) + o(\epsilon_n).$$

It results that $\epsilon_n \left(\partial_\epsilon E^\lambda(0, v^+) - \partial_\epsilon E^\lambda(0, v^*) + o(1)\right) \leq 0$ and we deduce that $\partial_\epsilon E^\lambda(0, v^+) \leq \partial_\epsilon E^\lambda(0, v^*)$.

Likewise, if $(v_n)_n$ converges to v^-, we get that $\partial_\epsilon E^\lambda(0, v^-) \leq \partial_\epsilon E^\lambda(0, v^*)$. $\qquad \square$

3.1.3. *Construction of the counterexample*

In conclusion, one needs to find a target, a well-chosen λ and v^* a global minimizer of $E^\lambda(0, \cdot)$ such that the inequality 3.18 is invalidated. There would consequently exist a deleted neighborhood of $\epsilon = 0$ (meaning a neighborhood of $\epsilon = 0$ without 0) for which there exists no continuous global minimizer of $E^\lambda(\epsilon, .)$. The sought-after vector fields v^{tar} and v^* must thus induce

$$\frac{\partial E^\lambda}{\partial \epsilon}(0, v^*) < \frac{\partial E^\lambda}{\partial \epsilon}(0, v^\alpha), \qquad (3.19)$$

where $v^\alpha \in \{v^+, v^-\}$. Let us recall that we chose a decreasing function ρ (which is the case of most usual kernels used to model varifolds) so that $\dot\rho(0) < 0$. Since all optimal curves have the same length, one can define $\ell_0 = \ell(v^*) = \ell(v^\alpha)$ and according to Lemma 3.10, this inequality 3.19 is equivalent to

$$\left(\ell(v^{\mathrm{tar}})-\ell_0\right)V(v^*)+\ell(v^{\mathrm{tar}})|x_{v^{\mathrm{tar}}}-x_{v^*}|^2 < \left(\ell(v^{\mathrm{tar}})-\ell_0\right)V(v^\alpha)+\ell(v^{\mathrm{tar}})|x_{v^{\mathrm{tar}}}-x_{v^\alpha}|^2.$$

Moreover, if we can have $\frac{4\lambda\ell(v^{\mathrm{tar}})}{4\lambda+1} > 2$ then $\ell_0 = \frac{4\lambda\ell(v^{\mathrm{tar}})}{4\lambda+1}$ and $\ell(v^{\mathrm{tar}}) - \ell_0 = 1/(4\lambda + 1)$. *In fine*, the counterexample must satisfy

$$\frac{V(v^*)}{4\lambda + 1} + \ell(v^{\mathrm{tar}})|x_{v^{\mathrm{tar}}} - x_{v^*}|^2 < \frac{V(v^\alpha)}{4\lambda + 1} + \ell(v^{\mathrm{tar}})|x_{v^{\mathrm{tar}}} - x_{v^\alpha}|^2. \quad (3.20)$$

Let us construct it explicitly. Consider for example $v^{\mathrm{tar}}(t) = a\mathbf{1}_{t\leq 1/2}$ with $a > 0$. The target curve $c_{v^{\mathrm{tar}}}$ is then given by $t \to (t, a(1/2 - t)^+)$ and we have $\ell(v^{\mathrm{tar}}) = (1 + \sqrt{a^2 + 1})$,

$$x_{v^{\mathrm{tar}}} = \left(0, \frac{\sqrt{a^2 + 1}}{1 + \sqrt{a^2 + 1}}\frac{a}{4}\right) \quad (3.21)$$

and

$$x_{v^\alpha} = \left(0, \frac{\alpha}{2}\sqrt{\frac{\ell^2}{4} - 1}\right), \quad (3.22)$$

where we assume that λ is large enough so that $\ell = \frac{4\lambda\ell(v^{\mathrm{tar}})}{4\lambda+1} > 2$.

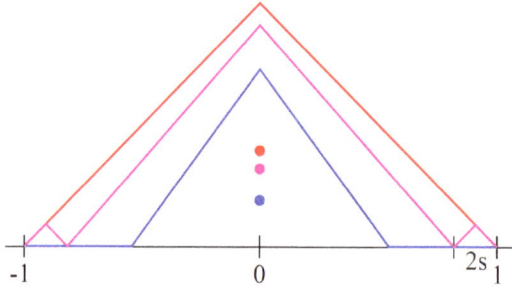

Fig. 7.   The target c_{tar} is the blue curve. The red curve is c_{v^+} where v^+ is the positive unique global continuous minimizer of $E^\lambda(0, v)$. The pink curve belongs to the set of curves generated by the $v^{s,*}$. The three dots in the middle are the respective centroid of the curves. One can see on this figure that $x_s^2 - x_{v^{\mathrm{tar}}}^2$ is strictly positive and it increases when s tends to 0 (the pink dot tends to the red dot when s tends to 0).

It results that the optimal continuous solution is $v^\alpha = v^+$. Let us introduce a set of vector fields $(v^{s,*})_{s\geq 0}$ defined by

$$v^{s,*}(t) = \sqrt{\ell^2/4 - 1}\left(1_{t<1-2s} + \mathrm{sign}(t - (1-s))1_{t\geq 1-2s}\right).$$

We have $v^+ = v^{0,*}$ and for any $s \geq 0$, $(v^{s,*})^2 + 1 \equiv \ell^2/4$ so that $v^{s,*}$ is a global minimizer of $E^\lambda(0,\cdot)$ that is not continuous when $s > 0$. In order to prove inequality 3.20, we just have to show that the derivative with respect to s of

$$s \mapsto \frac{V(v^{s,*})}{4\lambda + 1} + \ell(v^{\mathrm{tar}})|x_{v^{\mathrm{tar}}} - x_{v^{s,*}}|^2$$

is strictly negative on a neighborhood of $s = 0^+$.

Denote $x_s \doteq x_{v^{s,*}}$. We have

$$x_s = \left(0, \left(s^2 + (1-2s)(1-2s)/2\right)\sqrt{\ell^2/4 - 1}\right)$$

$$= \left(0, \left(3s^2 - 2s + \frac{1}{2}\right)\sqrt{\ell^2/4 - 1}\right).$$

One can easily show that $\frac{d}{ds}(|x_{v^{\mathrm{tar}}} - x_{v^{s,*}}|^2)_{|s=0} < 0$. It follows that $\frac{d}{ds}(V(v^{s,*}))_{|s=0} \leq 0$. If we denote $x_s = (x_s^1, x_s^2)$ then $s \mapsto x_s^1$ is constant and $\frac{dx_s^2}{ds}_{|s=0} < 0$. At last, we need to show that there exist a and λ such that $x_s^2 - x_{v^{\mathrm{tar}}}^2 > 0$. Assume then that λ is close to $+\infty$ so that $\ell = \ell(v^{\mathrm{tar}}) + o(1)$. Then since the sign of $g(a) = \sqrt{\ell^2/4 - 1}/2 - \frac{\sqrt{a^2+1}}{1+\sqrt{a^2+1}}a/4 = x_s^2 - x_{v^{\mathrm{tar}}}^2 + o(1)$ where $\ell = (1 + \sqrt{a^2 + 1})$ is strictly positive when $a > 0$ (see Figure 7), we deduce the final result.

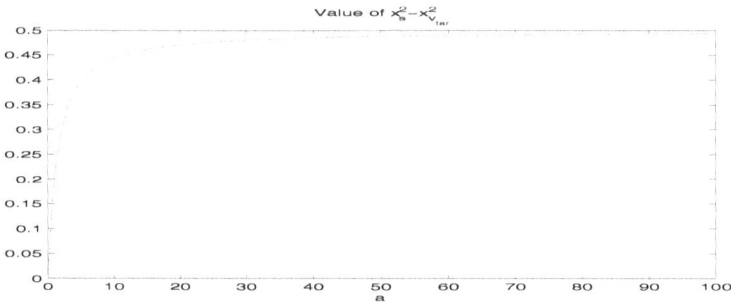

Fig. 8.   Plot of the function g.

In conclusion, we showed that for any $a > 0$, if λ is large enough and $\epsilon > 0$ small enough, the energy $E^\lambda(\epsilon,.)$ admits no global minimizer in

$\mathcal{C}([0,1], \mathbb{R}) \cap L^2([0,1], \mathbb{R})$. Let us remark additionally that this is not a consequence of the discontinuity of v^{tar}. Indeed, one can easily replace v^{tar} by an approximation in $\mathcal{C}^\infty$ with respect to the L^2-norm and deduce the same result.

Remark 3.12: Note that this counterexample could not be applied to the currents. Indeed, the choice of the kernel k_T is not open and the canceling effect of this kernel on opposite tangent vectors would reduce the length of the set of curves generated by the $v^{s,*}$ (the pink curve displayed in Figure 7).

3.2. Extension to the 3D case

As in the 2D case, we attempt now to show the following theorem for surfaces in $\mathbb{R}^3$.

Theorem 3.13: *There exist $v^{\text{tar}} \in L^2([0,1], \mathbb{R})$, $\lambda > 0$ and W such that E_W^λ has no time-continuous global minimizer.*

The main ideas of the proof remain the same. We consider as in Definition 3.2 a similar set of RKHS W_ϵ whose kernels are given by $k_\epsilon(x, y) = \rho(\epsilon|x - y|^2_{\mathbb{R}^3})$ where ρ is positive scalar function such that $\rho(0) = 1$, $\dot\rho$ is bounded, and $\dot\rho(0) < 0$. Proposition 3.15 will establish that $E_{W_0}^\lambda$ admits again exactly two continuous global minimizers v^+ and v^- among an infinite number of global minimizers. Proposition 3.17 will then show that the continuous solutions relative to $\epsilon > 0$ necessarily lie in a neighborhood of v^+ or v^-. At last, we will present a situation where the continuity is a constraint too restrictive as there exist global minimizers of $E_{W_0}^\lambda$ more stable with respect to ϵ than v^+ and v^-. In other words, if the energy increases more slowly around a discontinuous minimizer v^* than around v^+ and v^-, the existence of continuous global minimizers of $E_{W_\epsilon}^\lambda$ for ϵ in a deleted neighborhood of 0 is excluded. As before, this will require to compare the gradients of $E_{W_0}^\lambda$ with respect to ϵ at the minimizers of $v \mapsto E_{W_0}^\lambda(v)$. We denote again $E^\lambda(\epsilon, \cdot) = E_{W_\epsilon}^\lambda$.

The energy functions to minimize, associated to the spaces W_ϵ, are unchanged

$$E^\lambda(\epsilon, v) = \frac{1}{2} \int_0^1 v(t)^2 \, dt + \frac{\lambda}{2} |\mu_{v^{\text{tar}}} - \mu_v|^2_{W_\epsilon'}.$$

3.2.1. *Solutions for the degenerate kernel*

Consider now the case $\epsilon = 0$. The kernel of W_0 is the constant unit kernel. By analogy with the 2D case, the area of the surface γ_v is denoted $\ell(v)$ and we have

$$\ell(v) = |\mu_v|^2_{W_0'} = \int_0^{2\pi} \int_0^1 r\sqrt{1 + v(r)^2}\, dr d\theta \qquad (3.23)$$

$$= 2\pi \int_0^1 r\sqrt{1 + v(r)^2}\, dr\,. \qquad (3.24)$$

Remark 3.14: Note that for any $v \in L^2([0,1], V)$, we have $\ell(v) \geq \pi$. The growth process can only expand the initial unit disc.

The energy then reads

$$E^\lambda(0, v) = \frac{1}{2} \int_0^1 v(t)^2\, dt + \frac{\lambda}{2} \left(\ell(v^{\text{tar}}) - \ell(v)\right)^2\,.$$

The next proposition will establish the minimizers of this energy. For this purpose, given any constant $c \geq 1$, we will say that $v \in L^2([0,1], V)$ satisfies the (P_c) property if

$$(P_c) \quad \left| \begin{array}{l} \text{for almost any time } t \in [0,1], \\[4pt] v(t)^2 = \begin{cases} 0 & \text{if } t \leq \frac{1}{c} \\ (ct)^2 - 1 & \text{otherwise.} \end{cases} \end{array} \right.$$

Proposition 3.15: *For any $\lambda \geq 0$, there exists a unique constant $c_0 \geq 1$ such that:*
$v^* \in L^2([0,1], \mathbb{R})$ is a global minimizer of $E^\lambda(0, \cdot)$ if and only if it satisfies the (P_{c_0}) property.

Additionally, $c_0 = 1$ if and only if $\ell(v^{\text{tar}}) \leq \pi + 1/(2\pi\lambda)$. In this last case, $v^* \equiv 0$ is the unique global minimizer of $E^\lambda(0, \cdot)$.

Proof: The proof is similar as the one of Proposition 3.3. Introduce for $c \geq 1$

$$\rho_c(z, t) = \frac{z^2}{2} - ct\sqrt{z^2 + 1}\,,$$

defined on $\mathbb{R} \times [0,1]$. Given $t \in [0,1]$, the function $\rho_c(z, t)$ reaches its minimum at $z = 0$ if $t \leq \frac{1}{c}$ and at $z_c \doteq \pm \sqrt{(ct)^2 - 1}$ otherwise. Thus $v \in L^2([0,1], \mathbb{R})$ minimizes

$$\int_0^1 \rho_c(v(t), t)\, dt = \frac{1}{2} \int_0^1 v(t)^2\, dt - \frac{c}{2\pi} \ell(v)$$

if and only if it satisfies the (P_c) property. Now, if v_c satisfies (P_c) then

$$\ell(v_c) = 2\pi \int_0^1 t\sqrt{1 + v_c(t)^2}\, dt = 2\pi \left(\int_0^{\frac{1}{c}} t\, dt + \int_{\frac{1}{c}}^1 t\sqrt{(ct)^2}\, dt \right)$$

$$= 2\pi \left(\frac{1}{2c^2} + c \left[\frac{t^3}{3} \right]_{\frac{1}{c}}^1 \right) = \frac{2\pi}{3} c + \frac{\pi}{3} \frac{1}{c^2}.$$

Denote $\hat{\ell} : [1, +\infty[\to \mathbb{R}$ the function defined by

$$\hat{\ell}(c) = \frac{2\pi}{3} c + \frac{\pi}{3} \frac{1}{c^2} \tag{3.25}$$

and remark that $\hat{\ell}$ is a bijection from $[1, +\infty[$ to $[\pi, +\infty[$. (P_c) characterizes the minimizers of the constrained optimization problem

$$\left| \begin{array}{l} \min_{L^2} \int v(t)^2\, dt \\ \text{with } \ell(v) = \hat{\ell}(c) \end{array} \right. .$$

Therefore, (P_c) also determines exactly the minimizers of $E^\lambda(0, \cdot)$ when c minimizes

$$g(c) = E^\lambda(0, v_c) = \frac{1}{2} \int_{\frac{1}{c}}^1 (ct)^2 - 1\, dt + \frac{\lambda}{2} (\hat{\ell}(c) - \ell(v^{\text{tar}}))^2$$

$$= \frac{1}{2} \left(c^2 \left[\frac{t^3}{3} \right]_{\frac{1}{c}}^1 - \left(1 - \frac{1}{c} \right) \right) + \frac{\lambda}{2} \left(\frac{2\pi}{3} c + \frac{\pi}{3} \frac{1}{c^2} - \ell(v^{\text{tar}}) \right)^2$$

$$= \frac{c^2}{6} + \frac{1}{3c} - \frac{1}{2} + \frac{\lambda}{2} \left(\left(\frac{2\pi}{3} c + \pi \frac{1}{c^2} \right)^2 + \ell(v^{\text{tar}})^2 - 2\ell(v^{\text{tar}}) \left(\frac{2\pi}{3} c + \frac{\pi}{3} \frac{1}{c^2} \right) \right)^2$$

$$= \left(\frac{1}{6} + \frac{\lambda}{2} \left(\frac{2\pi}{3} \right)^2 \right) c^2 + C + \left(\frac{1}{3} + \frac{\lambda}{2} \left(\frac{2\pi}{3} \right)^2 \right) \frac{1}{c} + \frac{\lambda}{2} \left(\frac{\pi}{3} \right)^2 \frac{1}{c^4}$$

$$- \frac{\lambda}{2} \frac{2\pi}{3} \ell(v^{\text{tar}}) \left(2c + \frac{1}{c^2} \right),$$

where C is the constant $\frac{\lambda}{2} \ell(v^{\text{tar}})^2 - \frac{1}{2}$.
Since the uniqueness of c is required, let us study the variations of this function. We have

$$g'(c) = \left(\frac{1}{3} + \lambda \left(\frac{2\pi}{3} \right)^2 \right) c - \left(\frac{1}{3} + \frac{\lambda}{2} \left(\frac{2\pi}{3} \right)^2 \right) \frac{1}{c^2} - \frac{\lambda}{2} \left(\frac{2\pi}{3} \right)^2 \frac{1}{c^5}$$

$$+ \lambda \frac{2\pi}{3} \ell(v^{\text{tar}}) \left(\frac{1}{c^3} - 1 \right)$$

and

$$g''(c) = \left(\frac{1}{3} + \lambda \left(\frac{2\pi}{3}\right)^2\right) + 2\left(\frac{1}{3} + \frac{\lambda}{2}\left(\frac{2\pi}{3}\right)^2\right)\frac{1}{c^3} + 5\frac{\lambda}{2}\left(\frac{2\pi}{3}\right)^2\frac{1}{c^6} - 3\lambda\frac{2\pi}{3}\ell(v^{\mathrm{tar}})\frac{1}{c^4}.$$

For $c \geq 1$, $g'' = 0$ is thus equivalent to $h(c) = 0$ where $h(c) = c^4 g''(c)$. The derivative of h is given by

$$h'(c) = 4\left(\frac{1}{3} + \lambda\left(\frac{2\pi}{3}\right)^2\right)c^3 + 2\left(\frac{1}{3} + \frac{\lambda}{2}\left(\frac{2\pi}{3}\right)^2\right) - 10\frac{\lambda}{2}\left(\frac{2\pi}{3}\right)^2\frac{1}{c^3}$$
$$= c^{-3}Q(c^3),$$

where $Q(X) = 4\left(\frac{1}{3} + \lambda\left(\frac{2\pi}{3}\right)^2\right)X^2 + 2\left(\frac{1}{3} + \frac{\lambda}{2}\left(\frac{2\pi}{3}\right)^2\right)X - 10\frac{\lambda}{2}\left(\frac{2\pi}{3}\right)^2$.

Therefore, since Q is strictly increasing on $[1, +\infty[$ and $Q(1) = 2$, $h' > 0$ and h is strictly increasing on $[1, +\infty[$. Moreover, $\ell(v^{\mathrm{tar}}) \geq \pi$ so there exists $s \geq 1$ such that $\ell(v^{\mathrm{tar}}) = s\pi$. Then $h(1) = 1 + 2\pi^2\lambda(1 - s)$ and $h(1) < 0$ is equivalent to $s > 1 + 1/(2\pi^2\lambda)$. Under this condition, g'' has only one zero and g' is decreasing then increasing. Otherwise, g' is strictly increasing.

Finally, since $g'(1) = 0$, g has always only one global minimum on $[1, +\infty[$. Additionally, if $\ell(v^{\mathrm{tar}}) \leq \pi + 1/(2\pi\lambda)$, the minimizer is $c_0 = 1$ and corresponds to the solution $v^* \equiv 0$. Otherwise, $c_0 > 1$. $\square$

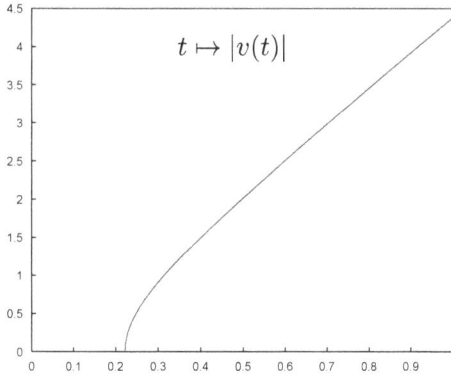

Fig. 9.   Plot of the norm of any optimal vector field. The (P_c) condition on this example is defined with $c = 4.5$ so that $\ell(v) \approx 3\pi$. The area of the surface has tripled with respect to its initial position.

Remark 3.16: As in the 2D case, the energy associated to the degenerate kernel admits two continuous global minimizers

$$v^+(t) \doteq \mathbb{1}_{t > \frac{1}{c_0}} \sqrt{(c_0 t)^2 - 1} \qquad \text{and} \qquad v^- \doteq -v^+. \qquad (3.26)$$

They are again surrounded by an infinite number of discontinuous global minimizers. However, these two solutions are not constant anymore. Indeed, in the 2D case, a constant vertical translation creates at all time the same amount of new matter measured by the length of the curve just created above the base between two times t and $t + \delta t$. In the 3D case, the surface created by a constant vertical translation between two times t and $t + \delta t$ is similar to a cylinder whose radius increases with t. The penalization term on v tends thus to accelerate the creation over time.

3.2.2. *Perturbation of the degenerate kernel*

As before, we will now follow the continuous global minimizers of $v \mapsto E^\lambda(\epsilon, v)$ when ϵ tends to 0 and show that they belong to a neighborhood of v^+ or v^-.

Proposition 3.17: *Assume that $\lambda > 0$, $\ell(v^{\mathrm{tar}}) > \pi + 1/(2\pi\lambda)$ and that for $\epsilon \geq 0$ small enough, there exists a global continuous minimum v_ϵ of $E^\lambda(\epsilon, \cdot)$. Then*

$$\lim_{\epsilon \to 0} \left(\min(|v_\epsilon - v^+|_\infty, |v_\epsilon - v^-|_\infty) \right) = 0 \qquad (3.27)$$

where v^+ and v^- are the only continuous global minimizers of $E^\lambda(0, \cdot)$.

Proof: We first show the convergence of the areas.

Lemma 3.18: *Denote $\ell_0 \doteq \ell(v^+) = \hat{\ell}(c_0)$, then $\ell(v_\epsilon)$ tends to ℓ_0 and $\ell_0 > \pi$.*

Proof: Consider the function $\hat{\ell}$ defined by equation 3.25. Recall that $\hat{\ell}$ is a bijection from $[1, +\infty[$ to $[\pi, +\infty[$ and as we said in Remark 3.14, that for any $v \in L_V^2$, $\ell(v) \geq \pi$. Therefore, for any $\epsilon \geq 0$, there exists a unique $c_\epsilon \geq 1$ such that $\ell(v_\epsilon) = \hat{\ell}(c_\epsilon)$. Let us show that

$$E^\lambda(0, v_\epsilon) \leq E^\lambda(\epsilon, v_\epsilon) + o(1) \leq E^\lambda(\epsilon, v^+) + o(1) \leq E^\lambda(0, v^+) + o(1).$$

Left and right inequalities result from the continuity of $E^\lambda(\cdot, v)$. Since v_ϵ minimizes $E^\lambda(\epsilon, \cdot)$, the central inequality is also true. Moreover, Proposition 3.15 ensures that for any v_{c_ϵ} that satisfies (P_{c_ϵ}), we also have $E^\lambda(0, v_{c_\epsilon}) \leq E^\lambda(0, v_\epsilon)$. We introduced in the proof of Proposition 3.15 a function g that satisfies for any $\epsilon \geq 0$, $g(c_\epsilon) = E^\lambda(0, v_{c_\epsilon})$. Moreover, c_0 is the unique minimum of g. It results that

$$g(c_0) \leq g(c_\epsilon) \leq E^\lambda(0, v_{c_\epsilon}) \leq E^\lambda(0, v^+) + o(1) \leq g(c_0) + o(1).$$

Hence, $g(c_\epsilon)$ tends to $g(c_0)$ and since g is continuous and increases around $+\infty$, c_ϵ tends to c_0. The continuity of $\hat\ell$ ensures at last that $\hat\ell(c_\epsilon)$ tends to $\hat\ell(c_0)$ so that $\ell(v_\epsilon)$ converges as announced to $\ell_0 = \hat\ell(c_0) = \ell(v^+)$. The last assertion is a direct result of Proposition 3.15 under the assumption that $\ell(v^{\text{tar}}) > \pi + 1/(2\pi\lambda)$. $\qquad\square$

Let us compute the gradient with respect to v of the energy

$$E^\lambda(\epsilon, v) = \frac{1}{2} \int_0^1 v(t)^2 \, dt + \frac{\lambda}{2} |\mu_v - \mu_{v^{\text{tar}}}|^2_{W'_\epsilon}.$$

Consider $\omega_v(\epsilon, \cdot) = K_{W_\epsilon}(\mu_v - \mu_{v^{\text{tar}}})$ given for any $x \in \mathbb{R}^3$ by

$$\omega_v(\epsilon, x) = K_{W_\epsilon}(\mu_v - \mu_{v^{\text{tar}}})(x) = \int_{\mathbb{R}^3} \rho(\epsilon |x - y|^2) \, d(\mu_v - \mu_{v^{\text{tar}}})(y),$$

so that

$$\mu_v(K_{W_\epsilon}(\mu_v - \mu_{v^{\text{tar}}})) = \int_0^{2\pi} \int_0^1 \omega(\epsilon, \gamma_v(\theta, r)) r \sqrt{1 + v(r)^2} \, dr d\theta.$$

We have then for any variation $\delta v \in L^2([0,1], \mathbb{R})$

$$\left(\partial_v E^\lambda(\epsilon, v) \,|\, \delta v \right) = \int_0^1 v(t) \, \delta v(t) \, dt + \lambda \left(\partial_v \mu_v \,|\, \delta v \right)(K_{W_\epsilon}(\mu_v - \mu_{v^{\text{tar}}}))$$

$$= \int_0^1 v(t) \, \delta v(t) \, dt + \lambda \int_0^{2\pi} \int_0^1 \left(\partial_2 \omega_v(\epsilon, \gamma_v(\theta, r)) \,|\, (0, 0, \int_r^1 \delta v_s \, ds) \right) r \sqrt{1 + v(r)^2}$$

$$+ \omega_v(\epsilon, \gamma_v(\theta, r)) \frac{r v(r) \delta v(r)}{\sqrt{1 + v(r)^2}} \, dr d\theta,$$

where $\partial_2 \omega_v(\epsilon, x) = 2\epsilon \int_{\mathbb{R}^3} \dot\rho(\epsilon |x - y|^2)(x - y) \, d(\mu_v - \mu_{v^{\text{tar}}})(y)$. Denote at last

$$z_{\epsilon,v}(r) \doteq \int_0^{2\pi} r \omega_v(\epsilon, \gamma_v(\theta, r)) \, d\theta,$$

$$\alpha_{\epsilon,v}(s) \doteq \lambda \int_0^{2\pi} \int_0^s \left(\partial_2 \omega_v(\epsilon, \gamma_v(\theta, r)) \,|\, (0, 0, 1) \right) r \sqrt{1 + v(r)^2} \, dr d\theta.$$

The gradient then reads

$$\nabla_v E^\lambda(\epsilon, v) = \left(1 + \lambda \frac{z_{\epsilon,v}}{\sqrt{1 + v^2}} \right) v + \alpha_{\epsilon,v} \qquad (3.28)$$

and for any $\epsilon \geq 0$, since v_ϵ is a zero of $\nabla_v E^\lambda(\epsilon, v)$, we have

$$\left(1 + \lambda \frac{z_{\epsilon, v_\epsilon}}{\sqrt{1 + v_\epsilon^2}}\right) v_\epsilon + \alpha_{\epsilon, v_\epsilon} = 0 \quad \text{a.e.} \, . \tag{3.29}$$

Now, on any bounded neighborhood of $(0,0)$ of $\mathbb{R}^+ \times L^2([0,1], V)$, γ_v is bounded, $d\mu_v$ and $d\mu_{v^{\text{tar}}}$ are finite, so that with $\dot{\rho}$ bounded we have $|\partial_2 \omega_v(\epsilon, \gamma_v)|_\infty = O(\epsilon)$ and thus

$$|\alpha_{\epsilon, v}|_\infty = O(\epsilon) \, .$$

Hence, for $\epsilon > 0$ small enough, there exist $M > 0$ and $\beta_\epsilon \geq 0$ such that we have for almost any $t \in [0,1]$ either

$$(i) \quad \left|\sqrt{1 + v_\epsilon(t)^2} - t\beta_\epsilon\right| \leq M\epsilon^{\frac{1}{2}}\sqrt{1 + v_\epsilon(t)^2} \quad \text{or} \quad (ii) \quad |v_\epsilon(t)| \leq M\epsilon^{\frac{1}{2}} \, .$$

Denote as before for any $\epsilon > 0$, $A_\epsilon \doteq \{t \in [0,1] \,|\, |\sqrt{v_\epsilon(t)^2 + 1} - t\beta_\epsilon| \leq M\epsilon^{\frac{1}{2}}\sqrt{v_\epsilon(t)^2 + 1}\}$. Lemma 3.18 says that $\ell(v_\epsilon)$ tends to ℓ_0 and $\ell_0 > \pi$. Moreover, if $v \equiv o(1)$, $\ell(v) = \pi + o(1)$. Hence, for ϵ small enough, $\lambda_\mathbb{R}(A_\epsilon) = 0$ implies then that $\ell(v_\epsilon) < \ell_0$ which is absurd. For any $t \in A_\epsilon$,

$$1 - M\epsilon^{\frac{1}{2}} \leq (1 - M\epsilon^{\frac{1}{2}})\sqrt{1 + v_\epsilon(t)^2} \leq t\beta_\epsilon \leq \beta_\epsilon \, ,$$

so that $\beta_\epsilon \geq 1 - M\epsilon^{\frac{1}{2}}$ and $A_\epsilon \subset [\frac{1 - M\epsilon^{\frac{1}{2}}}{\beta_\epsilon}, 1]$. Moreover, since $\lim_{\epsilon \to 0} z_{\epsilon, v}(t) = -2\pi t(\ell(v^{\text{tar}}) - \ell(v))$, Lemma 3.18 implies that $\beta_\epsilon = \beta_0 + o(1)$ where $\beta_0 = 2\pi\lambda(\ell(v^{\text{tar}}) - \ell_0)$.

Consider a small $\alpha > 0$ and denote $I_\alpha = [0, \frac{1}{\beta_0}(1 + \alpha)]$ and $I_\alpha^+ =]\frac{1}{\beta_0}(1 + \alpha), 1]$. Both (i) (restricted on $A_\epsilon \cap I_\alpha$) and (ii) imply that there exists $\eta > 0$ such that for any $\epsilon < \eta$ and almost any $t \in I_\alpha$,

$$|v_\epsilon(t)|^2 \leq 3\alpha \tag{3.30}$$

(using $1 + \alpha < \sqrt{1 + 3\alpha}$ for α small enough). Since v_ϵ is continuous, (v_ϵ) tends uniformly to 0 on $[0, \frac{1}{\beta_0}]$ (i.e. for any sequence $(v_{\epsilon_n})_n$ such that $\epsilon_n \to 0$). Let us show now that for $\epsilon > 0$ small enough, $I_\alpha^+ =]\frac{1}{\beta_0}(1 + \alpha), 1] \subset A_\epsilon$. There exists $\eta > 0$ such that for any $\epsilon < \eta$, we have

(1) for any $t \in I_\alpha^+$, $t\beta_\epsilon \geq 1 + \alpha/2$,
(2) $1 + M\epsilon^{\frac{1}{2}} \leq (1 + \alpha/2)(1 + \alpha/3)^{-\frac{1}{2}}$ and $M\epsilon^{\frac{1}{2}} \leq (\alpha/4)^{\frac{1}{2}}$.

Using $t\beta_\epsilon \leq (1 + M\epsilon^{\frac{1}{2}})\sqrt{1 + v_\epsilon(t)^2}$ when $t \in A_\epsilon$ and (ii) otherwise, we deduce that for any $\epsilon < \eta$ and any $t \in I_\alpha^+$, we have

$$\begin{cases} v_\epsilon(t)^2 \geq \frac{\alpha}{3} \text{ if } t \in A_\epsilon \, , \\ v_\epsilon(t)^2 \leq \frac{\alpha}{4} \text{ if } t \notin A_\epsilon \, . \end{cases}$$

Since v_ϵ is continuous, either $I_\alpha^+ \cap A_\epsilon$ or $I_\alpha^+ \cap A_\epsilon^c$ is empty. Since $\ell(v_\epsilon)$ tends to $\ell_0 > \pi$, it results that for any $\epsilon < \eta$, $I_\alpha^+ =]\frac{1}{\beta_0}(1+\alpha), 1] \subset A_\epsilon$.

Finally, v_ϵ converges uniformly to 0 on $[0, \frac{1}{\beta_0}]$ and $\sqrt{1 + v_\epsilon^2}$ converges uniformly to $t \mapsto \beta_0 t$ on any interval I_α^+ (for $\alpha > 0$) and with 3.30 we deduce that the convergence is uniform on $[0, 1]$. Since v_ϵ is continuous, the limit of any sequence $(v_{\epsilon_n})_n$ such that $\epsilon_n \to 0$ is also continuous and satisfies the (P_c) property for $c = \beta_0$. Additionally, equation 3.29 says that any minimizer of $E^\lambda(0, \cdot)$ must also satisfy (P_c). The uniqueness of c when (P_c) characterizes the minimizers of $E^\lambda(0, \cdot)$ (see Proposition 3.15) allows to conclude that there exists $M' > 0$ such that for any $\epsilon > 0$

$$|v_\epsilon - v^+|_\infty \le M'\epsilon \quad \text{or} \quad |v_\epsilon - v^+|_\infty \le M'\epsilon,$$

where v^+ and v^- are the two continuous global minimizers of $E^\lambda(0, \cdot)$. $\square$

As before, this proximity of continuous global minimizers induces some constraints on the slopes of the energy with respect to ϵ.

Proposition 3.19: *If there exists a decreasing sequence $\epsilon_n \to 0$ such that v_{ϵ_n} is a continuous global minimizers of $E^\lambda(\epsilon_n, \cdot)$ then for any global minimizers v^* of $E^\lambda(0, \cdot)$, we have*

$$\min \left(\partial_\epsilon E^\lambda(0, v^+), \partial_\epsilon E^\lambda(0, v^-) \right) \le \partial_\epsilon E^\lambda(0, v^*). \tag{3.31}$$

Proof: According to Proposition 3.17, there exists a subsequence of $(v_{\epsilon_n})_n$ which converges either to v^+ or v^-. The proof of Proposition 3.11 can then be applied here. $\square$

3.2.3. *Construction of the counterexample*

The final step is to exhibit an example for which inequality 3.31 does not occur. Recall the geometric expression of $\partial_\epsilon E^\lambda$ given by Lemma 3.10:

$$\partial_\epsilon E^\lambda(0, v) = \lambda \dot\rho(0) \left(\left(\ell(v^{\text{tar}}) - \ell(v) \right) \left(\ell(v^{\text{tar}}) V(v^{\text{tar}}) - \ell(v) V(v) \right) - \ell(v) \ell(v^{\text{tar}}) |x_{v^{\text{tar}}} - x_v|^2 \right),$$

where

$$x_v = \frac{1}{\ell(v)} \int x \, d\mu_v(x) \quad \text{and} \quad V(v) = \frac{1}{\ell(v)} \int |x - x_v|^2 \, d\mu_v(x).$$

Since all global minimizers of $E^\lambda(0, \cdot)$ have the same length, denote $\ell_0 = \ell(v^*) = \ell(v^+)$ and since $\dot\rho(0) < 0$, a counterexample should thus lead to a couple (v^*, v^+) satisfying:

$$V(v^*) \left[\ell(v^{\text{tar}}) - \ell_0 \right] + \ell(v^{\text{tar}}) |x_{v^{\text{tar}}} - x_{v^*}|^2 < V(v^+) \left[\ell(v^{\text{tar}}) - \ell_0 \right] + \ell(v^{\text{tar}}) |x_{v^{\text{tar}}} - x_{v^+}|^2.$$

$$\tag{3.32}$$

We exclude the negative continuous solution v^- as it is easy to show that for a target above the plane $Z = 0$, this solution will not be approached by any global minimizer of E_ϵ^λ for $\epsilon > 0$. Moreover, we have explicitly $V(v^+) = V(v^-)$ and if $\gamma_{v^{\mathrm{tar}}} \subset (Z \geq 0)$, $|x_{v^{\mathrm{tar}}} - x_{v^+}| < |x_{v^{\mathrm{tar}}} - x_{v^-}|$.

Proposition 3.20: *There exists a target such that for λ large enough inequality 3.32 occurs.*

As we saw earlier, the minimization of $E^\lambda(0, \cdot)$ admits either a unique solution (equal to 0) or an infinite number of solutions. In this last case, there are only two continuous solutions. One can observe that these solutions are those which, at a fixed area, produce the most widely deployed surface. We show with the following example that this property can be very restrictive. The partial derivative of E^λ with respect to ϵ at $(0, v)$, where v is a minimum, measures the stability of this minimum with respect to small variations of ϵ. Intuitively, the best candidate among this infinite number of solutions, is the one which generates the surface that is geometrically the closest to the target. The previous expression gives an explicit description of this closeness according to the attachment term we chose. It requires a small variance and a centroid close the target's one. Here is then a possible example.

The idea is to create a compact accordion in order to create a surface with a large area that yet remains close to the horizontal plane. Let us recall that

$$\gamma_v(\theta, r) = \left(r \cos \theta, r \sin \theta, \int_r^1 v_s \, ds\right)$$

and consider a target generated by one of the following vector fields

$$v_n^{\mathrm{tar}}(t) = nh s_n(t), \qquad \text{with} \qquad \begin{array}{ccc} [0,1] & \xrightarrow{s_n} & \{-1, 1\} \\ t & \mapsto & \mathbb{1}_{\lfloor nt \rfloor = 0[2]} - \mathbb{1}_{\lfloor nt \rfloor = 1[2]}, \end{array}$$

with $n \in \mathbb{N}$, $h \in]0, 1]$ a scale constant. Figure 10 displays an example for $n = 21$ and $h = 0.1$.

Denote z_v the third component of γ_v. It satisfies for $v = v_n^{\mathrm{tar}}$, for any $r \in [0, 1]$, $|z_{v_n^{\mathrm{tar}}}(r)| = |\int_r^1 v_n^{\mathrm{tar}}(s) \, ds| \leq h$. Moreover, $\ell(v_n^{\mathrm{tar}}) = 2\pi \sqrt{1 + (nh)^2}$. Therefore, no matter the choice of n, the target shape remains concentrated in $D \times [-h, +h]$ (where D is the unit disc). Yet, one can fix its area as large as necessary by increasing n.

The solutions v^* that minimize $E^\lambda(0, \cdot)$ are characterized by the (P_c) property with a optimal constant c to define and such that $\ell(v^*) = \hat{\ell}(c)$

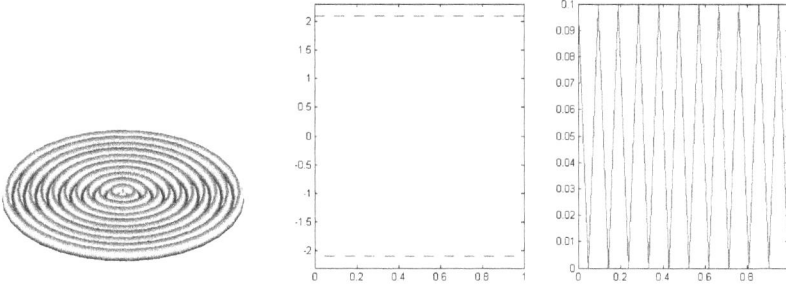

Fig. 10. From left to right: the target, the vector field which generates it from the unit disc with vertical translations, a radial cut of the surface (plot of the vertical component $z_{v^{\text{tar}}}(r)$). This example is essentially the 3D analogue of the 2D shape illustrated in Figure 4. For better visibility, colors indicate the height instead of the temporal tag.

denoted again ℓ_0. One can easily show that if λ tends to $+\infty$, ℓ_0 tends to $\ell(v^{\text{tar}})$. For λ large enough, we have thus

$$\ell_0 = \hat{\ell}(c) = \frac{2\pi}{3}c + \frac{\pi}{3c^2} \approx 2\pi\sqrt{1 + (nh)^2} = \ell(v^{\text{tar}}).$$

If hn is large enough, one can do the approximation $c \approx 3hn$.

Let us compare v^+ and v_n^* defined for any $t \in [0, 1]$ by

$$v^+(t) = \mathbb{1}_{t > \frac{1}{c}}\sqrt{(ct)^2 - 1} \qquad \text{and} \qquad v_n^*(t) = s_n(t)v^+(t),$$

where n is given by the choice of the target. They both satisfy (P_c). These two vector fields are displayed in Figure 12 and the surfaces that they generate are presented in Figure 11. When n increases, the continuous solution grows in space when the other one remains concentrated since $|z_{v_n^*}(r)| \leq \frac{1}{n}\sqrt{c^2 - 1} \approx 3h$.

Fig. 11. Surfaces generated from two solutions for the matching of the surface displayed in Figure 10. On the left: with the discontinuous vector field $s_n v^+$, on the right: with the continuous vector field v^+.

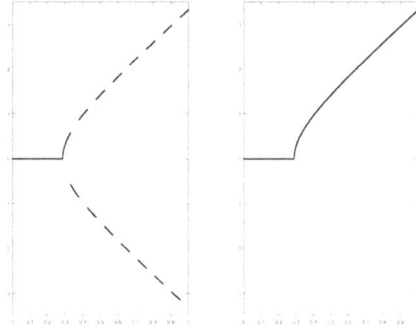

Fig. 12.   On the left: $v_n^* = s_n v^+$, on the right: v^+.

More precisely, for any surface generated by $v \in L^2$, the centroid belongs to the vertical axis through the origin. When x_{v^+} will move upwards when n increase, we have conversely for any n

$$|x_{v_n^{\text{tar}}}| \le \max_r |z_{v_n^{\text{tar}}}(r)| \le h \quad \text{and} \quad |x_{v_n^*}| \le \max_r |z_{v_n^*}(r)| \le 3h.$$

Likewise, for any $v \in L^2$

$$V(v) = \frac{1}{\ell(v)} \int_{\mathbb{R}^3} |x - x_v|^2 \, d\mu_v(x)$$

$$= \frac{1}{\ell(v)} \int (r \cos \theta)^2 + (r \sin \theta)^2 + |z_v(r) - x_v|^2 \, d\mu_v(x)$$

$$= \frac{1}{\ell(v)} \frac{2\pi}{3} + \frac{1}{\ell(v)} \int |z_v(r) - x_v|^2 \, d\mu_v(x).$$

It follows that

$$V(v_n^*) \le \frac{2\pi}{3\ell_0} + (6h)^2 \quad \text{and} \quad V(v^+) = \frac{2\pi}{3\ell_0} + \frac{1}{\ell_0} \int |z_{v^+}(r) - x_{v^+}|^2 \, d\mu_{v^+}(x).$$

In fine, if nh is fixed, $V(v^+)$ and $|x_{v^{\text{tar}}} - x_{v^+}|^2$ are fixed and strictly positive. Yet in the same time, if h tends to 0, $V(v_n^*)$ can be reduced to the minimal variance over the vector fields that satisfy (P_c) and $|x_{v_n^{\text{tar}}} - x_{v_n^*}|^2$ tends to 0. Therefore, the inequality

$$V(v^*) \left[\ell(v^{\text{tar}}) - \ell_0\right] + \ell(v^{\text{tar}})|x_{v^{\text{tar}}} - x_{v^*}|^2 < V(v^+) \left[\ell(v^{\text{tar}}) - \ell_0\right]$$
$$+ \ell(v^{\text{tar}})|x_{v^{\text{tar}}} - x_{v^+}|^2$$

can be satisfied.

In conclusion, note that $v^* = s_n v^+$ might not be the best candidate to minimize $\partial_\epsilon E^\lambda(\epsilon, \cdot)$ on a neighborhood of $\epsilon = 0$, but it was easy to demonstrate that it is strictly better than v^+ for n and λ large enough. As in the 2D case, one could generate similar surfaces with a smooth function s_n. This counterexample is not built on the discontinuity of v^{tar}.

At last, as pointed in Remark 3.16, this 3D example highlights a property of the optimal vector field that did not appear in the 2D case. With the growth dynamic, the norm of the optimal vector field tends to increase over time.

4. Existence of continuous minimizers in the current case

4.1. *Existence of global minimizers in* $L^2([0,1], V)$

In Section 2.3, we extended the current representation for any foliated shape generated by a control $v \in L^2([0,1], V)$ with the growth dynamic in the general situation of a coordinate space $X = [0,1] \times B$.

$$
\begin{aligned}
\left(L^2([0,1], V), |\cdot|_{L_V^2}\right) &\longrightarrow \quad \left(\mathcal{C}_0(\mathbb{R}^d, (\Lambda^k \mathbb{R}^d)^*), |\cdot|_\infty\right)^* \\
v &\longmapsto \mu_v : \omega \mapsto \int_0^1 \left[\int_{Y_t} \iota_{(h_t - v_t)} \phi_{t,1}^* \omega\right] dt ,
\end{aligned} \tag{4.1}
$$

where $(\phi_{s,t})_{s \leq t}$ is the flow of v, $\phi_{t,1}^* \omega$ is the pullback of ω by $\phi_{t,1}$, ι is the interior product and h_t is the unique vector field on $Y_t = q_0(B_t)$ defined for almost any $t \in [0,1]$ and any $x \in B_t$ by $h_t(q_0(x)) = \frac{\partial q_0}{\partial t}(x)$ where $B_t = \{t\} \times B \subset X$.

Unlike the varifolds, the currents provides a data attachment term that ensures the existence of continuous minimizers of

$$
E(v) \doteq \frac{1}{2} \int_0^1 |v|_V^2 \, dt + \frac{\lambda}{2} |\mu_{\text{tar}} - \mu_v|_{W'}^2 ,
$$

where μ_{tar} and μ_v are the currents associated to the target and the solution generated by v and W is now a RKHS embedded in the space of test functions $\mathcal{C}_0(\mathbb{R}^d, (\Lambda^k \mathbb{R}^d)^*)$. However, this result is not immediate. In this section, we first prove the existence of a solution in $L^2([0,1], V)$. The expression of the current μ_v 4.1 enlightens the foliation of our generated shapes and isolates each leaf. It allows to show a central property of current attachment terms that is not verified by varifold attachment terms: the lower semi-continuity (l.s.c.) on L_V^2.

Proposition 4.1: *For any* $\omega \in \mathcal{C}_0(\mathbb{R}^d, (\Lambda^k \mathbb{R}^d)^*)$, *the application* $v \to \mu_v(\omega)$ *is continuous with respect to the weak topology of* $L^2([0,1], V)$. *In particular,* $v \to |\mu_{\text{tar}} - \mu_v|_{W^*}^2$ *is l.s.c. with respect to the weak topology.*

Proof: We recall partially the assumptions on the space of vector fields V

$$(H_1^V) \quad \left| \quad \begin{array}{c} \text{There exists } c > 0 \text{ such that for any } v \in V \text{ and any } x \in \mathbb{R}^d, \\[4pt] |v(x)|_{\mathbb{R}^d} \leq c|v|_V(|x|_{\mathbb{R}^d} + 1). \end{array} \right.$$

$$(4.2)$$

Consider a weakly convergent sequence $v_n \rightharpoonup v_\infty$ in L_V^2, we have for any $\omega \in \mathcal{C}_0(\mathbb{R}^d, (\Lambda^k \mathbb{R}^d)^*)$

$$|\mu_{v^n}(\omega) - \mu_{v^\infty}(\omega)|$$
$$\leq \left| \int_0^1 \left[\int_{Y_t} \iota_{v_t^n - v_t^\infty} \phi_{t,1}^{v^\infty,*} \omega \right] dt \right| + \left| \int_0^1 \left[\int_{Y_t} \iota_{h_t - v_t^n} (\phi_{t,1}^{v^n,*} \omega - \phi_{t,1}^{v^\infty,*} \omega) \right] dt \right|.$$

$$(4.3)$$

The first term of the right-hand side is a continuous linear form ℓ on L_V^2 evaluated on $v^n - v^\infty$. *This is where the linearity of the currents attachment terms on the tangential data plays its role.* Indeed, we have for any $u \in L_V^2$

$$|\ell(u)| = \left| \int_0^1 \left[\int_{Y_t} \iota_{u_t} \phi_{t,1}^{v^\infty,*} \omega \right] dt \right|$$

$$\leq \sup_{t,y \in Y_t} |\phi_{t,1}^{v^\infty,*} \omega(y)|_\infty \left| \int_0^1 \left[\int_{Y_t} |u_t(y)|_{\mathbb{R}^d} \, d\mathcal{H}^{k-1}(y) \right] dt \right|$$

$$\leq \sup_{t,y \in Y_t} |d\phi_{t,1}^{v^\infty}|_\infty^k |\omega(y)|_\infty \left| \int_0^1 \left[\int_{Y_t} c(|y|_{\mathbb{R}^d} + 1)|u_t|_V \, d\mathcal{H}^{k-1}(y) \right] dt \right|$$

$$\leq c \sup_{t,y \in Y_t} |d\phi_{t,1}^{v^\infty}|_\infty^k |\omega(y)|_\infty \sup_{t,y \in Y_t} (|y|_{\mathbb{R}^d} + 1) \sup_t vol(Y_t) \int_0^1 |u_t|_V \, dt$$

$$\leq c'|u|_{L_V^2},$$

where $vol(Y_t)$ is the volume of Y_t. Consequently, since $(v^n)_n$ weakly converges to v^∞, $\ell(v^n - v^\infty)$ tends to 0. The second term can be bounded as follows

$$\left| \int_0^1 \left[\int_{Y_t} \iota_{h_t - v_t^n} (\phi_{t,1}^{v^n,*} \omega - \phi_{t,1}^{v^\infty,*} \omega) \right] dt \right| \leq m_1(n) m_2,$$

where $m_1(n) = \sup_{t,y \in Y_t} |\phi_{t,1}^{v^\infty,*} \omega(y) - \phi_{t,1}^{v^n,*} \omega(y)|_\infty$ tends to 0 and

$$m_2 = \sup_t vol(Y_t) \Big(\sup_X \left| \frac{\partial q_0}{\partial t}(x) \right| + c \sup_{t,y \in Y_t} (|y|_{\mathbb{R}^d} + 1) \underbrace{|v^n|_{L_V^2}}_{\leq \sup_n |v^n|_{L_V^2}} \Big).$$

We already know that if a sequence $(v_n)_n$ weakly converges to v_∞ then $(t,y) \to \phi_{t,1}^{v_n}(y)$ converges compactly to $(t,y) \to \phi_{t,1}^{v_\infty}(y)$. Moreover, since

$(v_n)_n$ is weakly convergent, $(v_n)_n$ is bounded, so that finally this upper bound tends to 0. Therefore, the function $v \mapsto \mu_v$, with values in $\mathcal{C}_0(\mathbb{R}^d, (\Lambda^k \mathbb{R}^d)^*)^*$, is continuous with respect to the weak topology of L^2_V and the first result is proved.

Moreover, since W is continuously embedded into $\mathcal{C}_0(\mathbb{R}^d, (\Lambda^k \mathbb{R}^d)^*)$, there exists $c' > 0$ such that for any linear form $\ell \in \mathcal{C}_0(\mathbb{R}^d, (\Lambda^k \mathbb{R}^d)^*)^*$, $|\ell(\omega)| \leq |\ell|_\infty |\omega|_\infty \leq c' |\ell|_\infty |\omega|_W$ so that $|\ell|_{W^*} \leq c' |\ell|_\infty$. It follows that for any $\omega \in W$, $\mu_{v^n}(\omega)$ tends to $\mu_{v^\infty}(\omega)$, i.e. μ_{v^n} weakly converges to μ_{v^∞} in W^*. Hence, $\mu_{v^n}(\mu_{\mathrm{tar}}) = \langle \mu_{v^n}, \mu_{\mathrm{tar}} \rangle_{W^*}$ tends to $\langle \mu_{v^\infty}, \mu_{\mathrm{tar}} \rangle_{W^*}$ and since the square norm of a Hilbert space is always lower semi-continuous with respect to the weak topology, we deduce that

$$
\begin{aligned}
|\mu_{\mathrm{tar}} - \mu_{v^\infty}|^2_{W^*} &= |\mu_{\mathrm{tar}}|^2_{W^*} - 2\langle \mu_{v^\infty}, \mu_{\mathrm{tar}} \rangle_{W^*} + |\mu_{v^\infty}|^2_{W^*} \\
&\leq |\mu_{\mathrm{tar}}|^2_{W^*} - 2\lim\langle \mu_{v^\infty}, \mu_{\mathrm{tar}} \rangle_{W^*} + \varliminf |\mu_{v^n}|^2_{W^*} \\
&\leq \varliminf \left(|\mu_{\mathrm{tar}}|^2_{W^*} - 2\langle \mu_{v^n}, \mu_{\mathrm{tar}} \rangle_{W^*} + |\mu_{v^n}|^2_{W^*} \right) \\
&\leq \varliminf |\mu_{\mathrm{tar}} - \mu_{v^n}|^2_{W^*}.
\end{aligned}
$$
□

This proposition induces a first main result: the existence of a solution *in $L^2([0,1], V)$* of the energy

$$
E(v) = \frac{1}{2} |v|^2_{L^2_V} + \frac{\lambda}{2} |\mu_{\mathrm{tar}} - \mu_v|^2_{W^*}.
$$

Theorem 4.2: *Consider $X = [0,1] \times B$ where B is a compact oriented manifold with corners and τ the projection on the first coordinate of X. Assume that $q_0 \in \mathcal{C}^\infty(X, \mathbb{R}^d)$. Consider the standard cost function*

$$
C(v) = \frac{1}{2} \int_0^1 |v_t|^2_V dt.
$$

Under the (H^{q_0}) and (H^V_1) conditions, the energy defined for any v in $L^2([0,1], V)$ by

$$
E(v) = C(v) + \frac{\lambda}{2} |\mu_v - \mu_{\mathrm{tar}}|^2_{W^*}
$$

admits a global minimizer.

Proof: Note that E is always positive. Let $(v^n)_n$ be a minimizing sequence of E. One can easily show that $(v^n)_n$ is bounded and we can then assume that v^n weakly converges in L^2_V. Denote v^∞ this limit. Proposition 4.1 says that E is lower semi-continuous with respect to the weak topology of L^2_V. It follows that $E(v^n)$ tends to $E(v^\infty)$ so that v^∞ minimizes E. □

Remark 4.3: One can generalize the previous theorem with a cost function C that satisfies $C(v)$ tends to $+\infty$ when $|v|_{L_V^2}$ tends to $+\infty$, e.g. for cost functions of the type

$$C(v) = \frac{1}{2} \int_0^1 \alpha_t |v_t|_V^2 dt,$$

where $\alpha : [0,1] \to \mathbb{R}^+$, as soon as α admits a strictly positive lower bound (see the so-called *adapted norm setup* [23]).

4.2. *Continuity of the global minimizers*

At this point, the continuity of a minimizer v^* of E is not acquired. We will prove now that all minimizers belong to $\mathcal{C}([0,1], V)$, which is not true when the attachment term is defined on varifolds. The outline of the proof is simple. We show that E is differentiable with respect to v and study the critical points of E. We keep the assumptions of the previous theorem. We assume in this section that W is a RKHS embedded in the space of $\mathcal{C}^1$ differential forms $\mathcal{C}_0^1(\mathbb{R}^d, (\bigwedge^k \mathbb{R}^d)^*)$.

We recall a standard result on the flow of a vector field.

Proposition 4.4: *Assume the* (H_1^V) *conditions given by equation 2.1. Let be* $v, \delta v \in L_V^2$ *and introduce the variations* $v^\epsilon = v + \epsilon \delta v$ *of* v *in the direction* δv *for* $\epsilon \in \mathbb{R}$. *Consider* $\phi_{s,t}^\epsilon$ *the flow of* v^ϵ, *meaning that* $\phi_{s,t}^\epsilon = \phi_t^\epsilon \circ \phi_s^{\epsilon,-1}$ *where* ϕ_t^ϵ *is the unique solution on* $[0,1]$ *of*

$$\phi_t^\epsilon = \mathrm{Id} + \int_0^t v_s^\epsilon \circ \phi_s^\epsilon \, ds.$$

Then, the application $\epsilon \to \left(\phi_{s,t}^\epsilon(y), d\phi_{s,t}^\epsilon(y)\right)$ *is of class* $\mathcal{C}^1$. *We have for any* $y \in \mathbb{R}^d$,

$$\frac{\partial}{\partial \epsilon} \phi_{s,t}^\epsilon(y)\Big|_{\epsilon=0} = \int_s^t d\phi_{u,t}(\phi_{s,u}(y)).\delta v_u(\phi_{s,u}(y)) \, du$$

and

$$\frac{\partial}{\partial \epsilon} d\phi_{s,t}^\epsilon(y)\Big|_{\epsilon=0} = \int_s^t \left[d^2\phi_{u,t}(\phi_{s,u}(y))d\phi_{s,u}(y)\right] \delta v_u(\phi_{s,u}(y))$$

$$+ d\phi_{u,t}(\phi_{s,u}(y))d\delta v_u(\phi_{s,u}(y))d\phi_{s,u}(y) \, du.$$

4.2.1. *Differentiability of the current representation*

A first step consists in studying the directional derivative of the current

$$\mu_v(\omega) = \int_0^1 \left[\int_{Y_t} \iota_{h_t - v_t} \phi_{t,1}^* \omega\right] dt$$

with respect to the vector field v. Let be $v, \delta v \in L_V^2$ and consider $v^\epsilon = v + \epsilon \delta v$ for $\epsilon \in \mathbb{R}$ and $\phi_{s,t}^\epsilon$ its flow. From the linearity of the interior product, we have

$$\mu_{v^\epsilon}(\omega) = \int_0^1 \left[\int_{Y_t} \iota_{h_t - v_t^\epsilon} \phi_{t,1}^{\epsilon,*} \omega \right] dt \tag{4.4}$$

$$= \int_0^1 \left[\int_{Y_t} \iota_{h_t - v_t} \phi_{t,1}^{\epsilon,*} \omega \right] dt - \epsilon \int_0^1 \left[\int_{Y_t} \iota_{\delta v_t} \phi_{t,1}^{\epsilon,*} \omega \right] dt . \tag{4.5}$$

We address the differentiation with respect to ϵ of these two terms separately. Denote

$$g(\epsilon) = \int_0^1 \left[\int_{Y_t} \iota_{h_t - v_t} \phi_{t,1}^{\epsilon,*} \omega \right] dt .$$

In order to rewrite g, let us introduce some notation. The variables are grouped in pairs:

$$\nu_s^\epsilon(y) = (v_s^\epsilon(y), dv_s^\epsilon(y)) \in \mathbb{R}^d \times \mathcal{L}(\mathbb{R}^d) ,$$
$$\varphi_s^\epsilon(y) = (\phi_{s,1}^\epsilon(y), d\phi_{s,1}^\epsilon(y)) \in \mathbb{R}^d \times \mathcal{L}(\mathbb{R}^d) ,$$
$$\delta\nu_s(y) = \frac{\partial}{\partial \epsilon} \nu_s^\epsilon(y)\big|_{\epsilon=0} = (\delta v_s(y), d\delta v_s(y)) ,$$
$$\delta\varphi_s(y) = \frac{\partial}{\partial \epsilon} \varphi_s^\epsilon(y)\big|_{\epsilon=0} . \tag{4.6}$$

Given $\omega \in W$, define $f_\omega : (\mathbb{R}^d \times \mathcal{L}(\mathbb{R}^d)) \to (\Lambda^k \mathbb{R}^d)^*$ such that $f_\omega(\varphi_t^\epsilon(y)) = (\phi_{t,1}^{\epsilon,*}\omega)_y$. This is, for any k-vector $\xi_1 \wedge \cdots \wedge \xi_k \in \Lambda^k \mathbb{R}^d$,

$$f_\omega(\varphi_t^\epsilon(y))(\xi_1 \wedge \cdots \wedge \xi_k) = \omega(\phi_{t,1}^\epsilon(y)) \left(d\phi_{t,1}^\epsilon(y)\xi_1 \wedge \cdots \wedge d\phi_{t,1}^\epsilon(y)\xi_k \right) .$$

We can easily check that f_ω is $\mathcal{C}^1$. At last, we get

$$g(\epsilon) = \int_0^1 \left[\int_{Y_t} \iota_{h_t - v_t} (f_\omega \circ \varphi_t^\epsilon) \right] dt .$$

Let us show now that g is derivable in 0 and let us explicit this derivative.

Lemma 4.5: *g is derivable and there exists $t \mapsto \mathcal{J}_t^a$ in $\mathcal{C}([0,1], V^*)$ such that*

$$g'(0) = \int_0^1 \mathcal{J}_s^a(\delta v_s) ds . \tag{4.7}$$

Proof: Denote $K = \bigcup_{s \in [0,1]} Y_s$, i.e. $K = q_0(X)$ and since $q_0 \in \mathcal{C}(X, \mathbb{R}^d)$ and X is compact, K is bounded. Proposition 4.4 implies that for any

$s \in [0,1]$, there exists $(t,y) \mapsto A_t^s(y)$ in $\mathcal{C}([0,1] \times \mathbb{R}^d, \mathcal{L}(\mathbb{R}^d \times \mathcal{L}(\mathbb{R}^d)))$ such that for any $y \in \mathbb{R}^d$, $\delta\varphi_s(y)$ 4.6 is given by

$$\delta\varphi_s(y) = \frac{\partial}{\partial\epsilon}\varphi_s^\epsilon(y)\big|_{\epsilon=0} = \int_s^1 A_t^s(y) \cdot \delta\nu_t(\phi_{s,t}(y)) dt. \tag{4.8}$$

We apply the Leibniz rule to derive under the integral sign so that we get

$$
\begin{aligned}
g'(0) &= \int_0^1 \left[\int_{Y_s} \iota_{h_s - v_s}(d_\varphi f_\omega(\varphi_s) \cdot \delta\varphi_s) \right] ds \\
&= \int_0^1 \left[\int_{Y_s} \iota_{h_s - v_s} d_\varphi f_\omega(\varphi_s) \int_s^1 A_t^s \cdot (\delta\nu_t \circ \phi_{s,t}) dt \right] ds \\
&= \int_0^1 \left[\int_{Y_s} \int_s^1 \iota_{h_s - v_s}\left((A_t^s)^* d_\varphi f_\omega(\varphi_s)\right) \cdot (\delta\nu_t \circ \phi_{s,t}) dt \right] ds,
\end{aligned}
$$

where for any $y \in \mathbb{R}^d$, $A_t^s(y)^*$ denotes the adjoint operator of $A_t^s(y) \in \mathcal{L}(\mathbb{R}^d, \mathcal{L}(\mathbb{R}^d))$. For any $y \in Y_s$, the integrand $\iota_{h_s - v_s}\left(A_t^s(y)^* \, d_\varphi f_\omega(\varphi_s(y))\right) \cdot (\delta\nu_t(\phi_{s,t}(y)))$ belongs to $(\Lambda^k\mathbb{R}^d)^*$ and we want to bound its norm independently of y to guarantee its integrability. This will come from the (H_1^V) conditions that gives a spatial control of the elements of V and their differential.

For any $y \in Y_s$, the application $A_t^s(y)^* \, d_\varphi f_\omega(\varphi_s(y))$ belongs to $\mathcal{L}\left(\mathbb{R}^d \times \mathcal{L}(\mathbb{R}^d), (\Lambda^k\mathbb{R}^d)^*\right)$ and can be identified to an element of $(\Lambda^k\mathbb{R}^d)^* \otimes (\mathbb{R}^d \times \mathcal{L}(\mathbb{R}^d))^*$. Moreover, for any $\zeta \in (\Lambda^k\mathbb{R}^d)^* \otimes (\mathbb{R}^d \times \mathcal{L}(\mathbb{R}^d))^*$, consider $l_y^\zeta : V \to (\Lambda^k\mathbb{R}^d)^*$ by

$$l_y^\zeta(u) = \zeta\left(u(y), du(y)\right).$$

Then l_y^ζ is linear and under the (H_1^V) conditions, there exists $c_V \in V$, such that for any $u \in V$, for any $y \in K$, if $\mu = (u, du)$, then

$$
\begin{aligned}
|l_y^\zeta(u)|_{(\Lambda^k\mathbb{R}^d)^*} &= |\zeta(\mu(y))|_{(\Lambda^k\mathbb{R}^d)^*} \\
&\leq |\zeta|_{(\Lambda^k\mathbb{R}^d)^* \otimes (\mathbb{R}^d \times \mathcal{L}(\mathbb{R}^d))^*} |\mu(y)|_{\mathbb{R}^d \times \mathcal{L}(\mathbb{R}^d)} \\
&\leq c_V|u|_V \sup_{y \in K}(1 + |y|_{\mathbb{R}^d})|\zeta|_{(\Lambda^k\mathbb{R}^d)^* \otimes (\mathbb{R}^d \times \mathcal{L}(\mathbb{R}^d))^*} .
\end{aligned}
$$

Hence, l_y^ζ belongs to $(\Lambda^k\mathbb{R}^d)^* \otimes V^*$ and

$$|l_y^\zeta|_{(\Lambda^k\mathbb{R}^d)^* \otimes V^*} \leq c_V \sup_{y \in K}(1 + |y|_{\mathbb{R}^d})|\zeta|_{(\Lambda^k\mathbb{R}^d)^* \otimes (\mathbb{R}^d \times \mathcal{L}(\mathbb{R}^d))^*} .$$

We can therefore apply Fubini's theorem to get that for any $u \in V$

$$\mathcal{J}_t^a(u) \doteq \int_0^t \left(\int_{Y_s} \iota_{h_s - v_s} \left(l_{\phi_{s,t}}^{(A_t^s)^* d_\varphi f_\omega(\varphi_s)}(u) \right) \right) ds. \tag{4.9}$$

Finally, since $t \to A_t^s(y)$ is continuous, we deduce easily that $t \mapsto \mathcal{J}_t^a$ is continuous. □

To study the second term of μ^ϵ in equation 4.5, we introduce the next lemma.

Lemma 4.6: *Given* $\omega \in \mathcal{C}_0^1(\mathbb{R}^d, (\Lambda^k \mathbb{R}^d)^*)$, *define for any* $t \in [0,1]$ *and any* $u \in V$

$$\mathcal{J}_t^b(u) = \int_{Y_t} \iota_u(\omega). \tag{4.10}$$

Then $\mathcal{J}^b$ *belongs to* $\mathcal{C}([0,1], V^*)$.

Proof: For any $t \in [0,1]$, $\mathcal{J}_t^b$ is linear on V and with the (H_1^V) conditions, there exists $c > 0$ such that for any $u \in V$,

$$|\mathcal{J}_t^b(u)| \leq \int_{Y_t} |u(y)|_{\mathbb{R}^d} |\omega(y)|_\infty d\mathcal{H}^{k-1}(y)$$
$$\leq c \sup_{y \in K} (1 + |y|_{\mathbb{R}^d})\, vol(Y_t) |\omega|_\infty |u|_V .$$

Thus, $\mathcal{J}_t^b \in V^*$. Moreover, we can show that $t \to \mathcal{J}_t^b$ is differentiable (and in particular continuous). From the spatial regularity of any $u \in V$, we deduce that $\omega^u \doteq \iota_u(\omega) \in \mathcal{C}^1(\mathbb{R}^d, (\Lambda^{k-1}\mathbb{R}^d)^*)$. Now, under the (H^{q_0}) conditions, we can pull backward the integrand of $\mathcal{J}^b$:

$$\mathcal{J}_t^b(u) = \int_{\{t\} \times B} q_0^* \omega^u .$$

Therefore, if $\frac{\partial}{\partial t}$ is the vector field on X defined at any point $(t, x_B) \in [0,1] \times B$ by $(1, 0_{T_{x_B} B})$, then $\frac{\partial}{\partial t}$ generates a flow ψ_t on X satisfying $\psi_t(s, x_B) = (s+t, x_B)$. Thus, $\alpha \doteq q_0^* \omega^u \in \mathcal{C}^1(X, \Lambda^{k-1} T^* X)$ is a $(k-1)$-form on X and it results from Cartan's formula and Stokes' theorem (see Corollary 6.4) that

$$\frac{d}{dt} \mathcal{J}_t^b(u) = \frac{d}{dt} \int_{\{t\} \times B} \alpha = \int_{\{t\} \times B} \iota_{\frac{\partial}{\partial t}} d\alpha + \int_{\{t\} \times \partial B} \iota_{\frac{\partial}{\partial t}} \alpha . \tag{4.11}$$

□

4.2.2. *Continuity of the minimizers*

Finally, we can conclude that all solutions are continuous and the next theorem recalls all assumptions.

Theorem 4.7: *Consider* $X = [0,1] \times B$ *where* B *is a compact oriented manifold with corners and* τ *the projection on the first coordinate of* X.

Assume that $q_0 \in \mathcal{C}^\infty(X, \mathbb{R}^d)$. Under the (H^{q_0}) and (H_1^V) conditions, if $v^* \in L^2([0,1], V)$ minimizes the energy defined by

$$E(v) = \frac{1}{2} \int_0^1 |v|_V^2 dt + \frac{\lambda}{2} |\mu_v - \mu_{\mathrm{tar}}|_{W^*}^2 \,,$$

then v^* belongs to $\mathcal{C}([0,1], V)$.

More precisely, for any $(v, \delta v)$ in $L_V^2 \times L_V^2$, the application $\epsilon \mapsto g(\epsilon) \doteq E(v + \epsilon \delta v)$ is differentiable at 0 and we have

$$g'(0) = \int_0^1 \langle v_t, \delta v_t \rangle dt + \lambda \int_0^1 \left[\int_{Y_t} \iota_{h_t - v_t} \left(\frac{\partial \phi_{t,1}^* \omega}{\partial v} . \delta v \right) \right] dt - \lambda \int_0^1 \left[\int_{Y_t} \iota_{\delta v_t} \phi_{t,1}^* \omega \right] dt$$

$$= \int_0^1 L_V v_t(\delta v_t) + \mathcal{J}_t^a(\delta v_t) - \mathcal{J}_t^b(\delta v_t) \, dt \,,$$

where K_V and $L_V = K_V^{-1}$ are the isomorphisms between V and V^*, $\omega = K_W(\mu_v - \mu_{tar})$, $\mathcal{J}^a, \mathcal{J}^b \in \mathcal{C}([0,1], V^*)$ are defined by equations 4.9 and 4.10 and h_t is the unique vector field on Y_t defined for almost any $t \in [0,1]$ and any $x \in B_t$ by $h_t(q_0(x)) = \frac{\partial q_0}{\partial t}(x)$.

Proof: We have

$$\frac{\partial}{\partial \epsilon} \frac{1}{2} |\mu_v - \mu_{\mathrm{tar}}|_{W^*}^2 \Big|_{\epsilon=0} = \frac{\partial}{\partial \epsilon} \mu_{v^\epsilon}(\omega) \Big|_{\epsilon=0} \,.$$

The expression of $\mu_{v^\epsilon}(\omega)$ is given by equation 4.5 and its derivative with respect to ϵ is given above in Section 4.2.1.

At last, if v^* minimizes E then $L_V v_t^* = \mathcal{J}_t^b - \mathcal{J}_t^a$ for almost every $t \in [0,1]$. Since $\mathcal{J}^a$ and $\mathcal{J}^b$ are continuous, $t \mapsto v_t^* = K_V(\mathcal{J}_t^b - \mathcal{J}_t^a)$ is continuous at any $t \in [0,1]$. □

Remark 4.8: One can easily generalize this theorem with a cost function on L_V^2 of the type $C(v) = \frac{1}{2} \int_0^1 C(v_t, t) dt$. More precisely, assume that there exists $\ell \in \mathcal{C}([0,1], V^*)$ such that for any $t \in [0,1]$, $\frac{\partial C}{\partial v}(v, t) = \ell_t(v)$ and ℓ_t is invertible. If $v^* \in L^2([0,1], V)$ minimizes the energy

$$E(v) = \frac{1}{2} \int_0^1 C(v_t, t) dt + \frac{\lambda}{2} |\mu_v - \mu_{\mathrm{tar}}|_{W^*}^2 \,,$$

then for any $t \in [0,1]$

$$v_t^* = \ell_t^{-1} \big(K_V(\mathcal{J}_t^b - \mathcal{J}_t^a) \big) \,.$$

It follows that $v^* \in \mathcal{C}([0,1], V)$.

Remark 4.9: Note that $\mathcal{J}_0^a$ is always null. Moreover, if $\mathcal{H}^{k-1}(Y_t)$ is null, then $\mathcal{J}_t^b$ is also null. In the case of the horn, Y_0 represents the tip of the horn. It is thus reduced to a point, so that v_0^* is necessarily vanishing.

5. Conclusion

We examined in this paper a growth process by foliation. In the large class of growth mapped evolutions, the existence of a foliation induced by the birth tag guaranties some regularity of the growth process. Each image growth mapped evolution of a given biological coordinate system inherits its foliation that is then a key element to describe and to overcome the lack of spatial regularity of the generated shapes and to define current and varifold representations.

The growth dynamic is the first time-varying dynamic introduced for the analysis of longitudinal shape data in the context of shape space. We studied the existence and continuity of global minimizers v of the optimization problem for the assimilation of time-varying shapes in the specific case of the growth dynamic. These questions lie on the choice of the data attachment term. We exhibited two counterexamples for the varifold representation. These situations highlighted the lack of spatial regularity of a shape generated by a discontinuous time-varying vector field $t \mapsto v_t$. This issue is well addressed by the current representation that has a regularization effect on the shapes. We proved indeed, with a data attachment term built on a current representation, the existence of global minimizers as well as their continuity.

6. Annex: Reminder on differential geometry

A k-dimensional *manifold with corners* extends the definition of *regular* manifolds (in the usual sense) to allow the shape to locally resemble a semi-orthant of $\mathbb{R}^k$. At any $x_0 \in X$, there exists a chart (U, ψ)

$$
\begin{aligned}
U &\longrightarrow \mathbb{R}^{k-p} \times \mathbb{R}^p_+ \\
x &\mapsto (x_1, \ldots, x_{k-p}, y_1, \ldots, y_p),
\end{aligned}
\tag{6.1}
$$

centered at x_0, i.e. $\psi(x_0) = (0, \cdots, 0)$, between a open set $U \ni x_0$ and a semi-orthant $\mathbb{R}^{k-p} \times \mathbb{R}^p_+$ for an integer $p = p(x_0) \geq 0$. For a regular manifold, p is always null. If p takes values only in $\{0, 1\}$ on X, then X is called a *manifold with boundary*.

We denote $\mathcal{H}^k$ the k-dimensional **Hausdorff measure** on $\mathbb{R}^d$. We remind that $\mathcal{H}^k$ is defined as an outer measure on $\mathbb{R}^d$ that basically measures the k-dimensional volume of a subset of $\mathbb{R}^d$. In particular, when $k = d$, we have $\mathcal{H}^d = \lambda^d$ the usual Lebesgue measure. If M is a p-dimensional submanifold of $\mathbb{R}^d$, then $\mathcal{H}^k(M)$ is the k-volume of M if $p = k$, vanishes if $p < k$ and equals $+\infty$ when $k < p$.

The interior product exhibits in Section 4.1 the linearity property of the currents with respect to the tangential data.

Definition 6.1: The interior product is defined to be the contraction of a differential form with a vector field. Thus if v is a vector field on the manifold M, then

$$\iota_v : (\Lambda^k M)^* \to (\Lambda^{k-1} M)^*$$

is the map which sends a k-form ω to the $(k-1)$-form $\iota_v \omega$ defined by the property that for any $m \in M$, $(k-1)$-vector $\xi_1 \wedge \cdots \wedge \xi_{k-1}$, $\xi_i \in T_m M$,

$$(\iota_v \omega)(m)\big(\xi_1 \wedge \cdots \wedge \xi_{k-1}\big) \doteq \omega(m)\big(v(m) \wedge \xi_1 \wedge \cdots \wedge \xi_{k-1}\big) \,.$$

Hence, ι is linear with respect to v.

Corollary 6.4, given hereafter, results from Stokes' theorem and Cartan's formula and plays a central role to exploit the linearity of the current representation with respect to the tangential data of a shape. It is used in Lemma 4.6.

Theorem 6.2 (Stokes' theorem). *Let M be an oriented compact k-dimensional differential manifold with corners. For any differential $(k-1)$-form ω of class $\mathcal{C}^1$*

$$\int_M d\omega = \int_{\partial M} \omega \,.$$

Proof: See [28]. $\qquad\qquad\qquad\qquad\qquad\qquad\qquad\qquad\qquad\qquad\qquad\qquad$ □

The Lie derivative of differential forms with respect to vector fields in the direction of a vector field v expresses how a current associated to a shape X varies when X is deformed in the direction of v. More precisely, given a flow ϕ_t such that $\phi_0 = \mathrm{Id}$ and $\dot\phi_{t|t=0} = v$

$$\mathcal{L}_v \omega = \lim_{t \to 0} \frac{\phi_t^* \omega - \omega}{t} = \frac{\partial}{\partial t} \phi_t^* \omega_{|t=0} \,. \tag{6.2}$$

It follows that

$$\frac{\partial}{\partial t} \mu_{\phi_t(X)}(\omega)_{|t=0} = \mu_X(\mathcal{L}_v \omega) \,. \tag{6.3}$$

Theorem 6.3 (Cartan's formula). *Let ω be a differential form of class $\mathcal{C}^1$ and v a vector field then*

$$\mathcal{L}_v \omega = d\iota_v \omega + \iota_v d\omega \,.$$

Proof: See [27] Lemma 7.2.1 and 10.3.2. □

We apply Cartan's formula in a particularly simple case. The manifold M is embedded in $[0,1] \times M$ and the deformation is the translation along the first coordinate.

Corollary 6.4: *Denote $v = \partial_t$ the vector field defined at any point $(t,m) \in [0,1] \times M$ by $(1, 0_{T_m M})$ and $M_t = \{t\} \times M$ then*

$$\frac{\partial}{\partial t} \left(\int_{M_t} \omega \right) \Big|_{t=0} = \int_{\partial M} \iota_v \omega + \int_M \iota_v d\omega .$$

Proof: We deduce from Cartan's formula that

$$\frac{\partial}{\partial t} \left(\int_{M_t} \omega \right) \Big|_{t=0} = \frac{\partial}{\partial t} \int_M \phi_t^* \omega|_{t=0} = \int_M \mathcal{L}_v \omega = \int_M d\iota_v \omega + \iota_v d\omega ,$$

where ϕ_t is the local flow resulting from v. Stokes' theorem allows then to conclude. □

References

1. S. Allassonniere, A. Trouvé, and L. Younes. Geodesic shooting and diffeomorphic matching via textured meshes. *EMMCVPR05*, 365–381, 2005.
2. S. Arguillère, E. Trélat A. Trouvé and L. Younes. Shape Deformation Analysis From the Optimal Control Viewpoint. *Journal de Mathématiques Pures et Appliquées*, Volume 104, Issue 1, July 2015, Pages 139–178.
3. S. Arguillère. The abstract setting for shape deformation analysis and lddmm methods. In *Geometric Science of Information*, pages 159–167. Springer, 2015.
4. S. Arguillère. The general setting for shape analysis. *arXiv preprint arXiv:1504.01767*, 2015.
5. B. B. Avants, C. L. Epstein, M. Grossman, and J. C. Gee. Symmetric diffeomorphic image registration with cross-correlation: evaluating automated labeling of elderly and neurodegenerate brain. *Medical image analysis*, 12(1):26–41, 2008.
6. M.F. Beg, M.I. Miller, A. Trouvé and L. Younes. Computing large deformation metric mappings via geodesic flows of diffeomorphisms. *International journal of computer vision*, 61 (2), 139–157, 2005.
7. M. Bauer, M. Bruveris, P. Harms, and P. W. Michor. Geodesic distance for right invariant sobolev metrics of fractional order on the diffeomorphism group. *Annals of Global Analysis and Geometry*, 44(1):5–21, 2013.
8. N. Charon and A. Trouvé. The Varifold Representation of Non-oriented Shapes for Diffeomorphic Rregistration. *SIAM journal on Imaging Sciences*, 6:2547–2580, 2013.

9. S. Durrleman, M. Prastawa, N. Charon, J. R. Korenberg, S. Joshi, G. Gerig and A. Trouvé. Morphometry of anatomical shape complexes with dense deformations and sparse parameters. *Neuroimage.* 2014 Nov 1;101:35–49. doi: 10.1016/j.neuroimage.2014.06.043. Epub 2014 Jun 26.
10. S. Durrleman, X. Pennec, A. Trouvé, J. Braga, G. Gerig, and N. Ayache. Toward a comprehensive framework for the spatiotemporal statistical analysis of longitudinal shape data. *Int. Journal of Computer Vision*, 103(1):22–59, 2013.
11. S. Durrleman, P. Fillard, X. Pennec, A. Trouvé, and N. Ayache. Registration, atlas estimation and variability analysis of white matter fiber bundles modeled as currents. *NeuroImage*, 2010.
12. S. Durrleman, X. Pennec, A. Trouvé, N. Ayache, and J. Braga. Comparison of the endocast growth of chimpanzees and bonobos via temporal regression and spatiotemporal registration, *1st workshop on Spatiotemporal Image Analysis for Longitudinal and Time-Series Image Data*, Beijing, Septembre 2010.
13. J. Feydy, B. Charlier, F.-X. Vialard, G. Peyré. Optimal Transport for Diffeomorphic Registration. *MICCAI 2017.*, Sep 2017, Quebec, Canada. 2017, Proc. MICCAI 2017.
14. P. T. Fletcher. Geodesic regression and the theory of least squares on riemannian manifolds. *International journal of computer vision*, 105(2):171–185, 2013.
15. J. Glaunès. *Transport par difféomorphismes de points, de mesures et de courants pour la comparaison de formes et l'anatomie numérique*. PhD thesis, 2005.
16. J. Glaunès, A. Trouvé, and L. Younes. Diffeomorphic matching of distributions: A new approach for unlabelled point-sets and sub-manifolds matching. *IEEE Computer Society Conference on Computer Vision and Pattern Recognition*, 2:712–718, 2004.
17. J. Glaunès, A. Qiu, M. Miller, and L. Younes. Large deformation diffeomorphic metric curve mapping. *Int J Comput Vis*, 80(3):317–336, 2008.
18. U. Grenander. General pattern theory: A mathematical study of regular structures. *Clarendon Press Oxford*, 1993.
19. U. Grenander and M. I. Miller. Computational anatomy: an emerging discipline. *Quarterly of Applied Mathematics - Special issue on current and future challenges in the applications of mathematics*, Volume LVI Issue 4, 617–694, 1998.
20. U. Grenander, A. Srivastava, and S. Saini. *Characterization of biological growth using iterated diffeomorphisms*. 3rd IEEE International Symposium on Biomedical Imaging: Nano to Macro, 2006.
21. K. Ito and K. Kunisch. Lagrange Multiplier Approach to Variational Problems and Applications. volume 15 of *Advances in Design and Control*, Society for Industrial and Applied Mathematics (SIAM), Philadelphia, PA, 2008.
22. S.C. Joshi and M.I. Miller. Landmark matching via large deformation diffeomorphisms. *IEEE Transcript Image Processing*, 9(8):1357–1370, 2000.
23. I. Kaltenmark Geometrical Growth Models for Computational Anatomy. *PhD Thesis*, ENS Paris-Saclay, 2016.

24. I. Kaltenmark and A. Trouvé Estimation of a Growth Development with Partial Diffeomorphic Mappings. *Preprint.*

25. I. Kaltenmark and A. Trouvé Partial Matchings and Growth Mapped Evolutions in Shape Spaces. *The IEEE Conference on Computer Vision and Pattern Recognition (CVPR) Workshops*, pp. 28–36, 2016.

26. I. Kaltenmark, B. Charlier and N. Charon A general framework for curve and surface comparison and registration with oriented varifolds. *Computer Vision and Pattern Recognition (CVPR)*, 2017.

27. F. Labourie. Geométrie différentielle. 2013. Available at http://math.unice.fr/~labourie/preprints/pdf/geomdiff.pdf.

28. J.M. Lee. Introduction to Smooth Manifolds (Second Edition). *Springer*, 2016.

29. D. Lombardi and E. Maitre. Eulerian models and algorithms for unbalanced optimal transport. *hal-00976501v3*, 2013.

30. P.W. Michor and D. Mumford. An overview of the Riemannian metrics on spaces of curves using the Hamiltonian approach. *Appl. Comput. Harmon. Anal.*, 23(1):74–113, 2007.

31. M.I. Miller, A. Trouvé, and L. Younes. On the metrics and euler-lagrange equations of computational anatomy. *Annual Review of Biomedical Engineering*, 4:375–405, 2002.

32. M.I. Miller, L. Younes, and A. Trouvé. Diffeomorphometry and geodesic positioning systems for human anatomy. *Technology (Singap World Sci)*, 02(01), 36–43, 2014.

33. M.I. Miller, A. Trouvé and L. Younes. Hamiltonian Systems and Optimal Control in Computational Anatomy: 100 Years Since D'Arcy Thompson. *Annual Review of Biomedical Engineering*, Vol. 17, 447–509, 2015.

34. M.I. Miller, L. Younes, J.T. Ratnanather, T. Brown, H. Trinh, D.S. Lee, D. Tward, P.B. Mahon, S. Mori, M. Albert, BIOCARD Research Team. Amygdalar atrophy in symptomatic Alzheimer's disease based on diffeomorphometry: the BIOCARD cohort. *Neurobiology of aging*, Vol. 36, S3–S10, 2015.

35. M. Niethammer, Y. Huang, and F.-X. Vialard. Geodesic Regression for Image Time-Series. *Medical Image Computing and Computer-Assisted Intervention*, 14(Pt 2):655–662, 2011.

36. B. Piccoli and F. Rossi. Generalized Wasserstein distance and its application to transport equations with source. *Archive for Rational Mechanics and Analysis*, 211(1):335–358, 2014.

37. A. Qiu, L. Younes, M. I. Miller, and J. G. Csernansky. Parallel transport in diffeomorphisms distinguishes the time-dependent pattern of hippocampal surface deformation due to healthy aging and the dementia of the alzheimer's type. *NeuroImage*, 40(1):68–76, 2008.

38. G. Reeb. *Sur certaines propriétés topologiques des variétés feuilletées*. Hermann, 1952.

39. P. Roussillon and J. Glaunès. Kernel Metrics on Normal Cycles and Application to Curve Matching. *SIAM Journal on Imaging Sciences*, 9(4):1991–2038, 2016.

40. S. Schwabik and Y. Guoju. *Topics in Banach Space Integration*. World Scientific, 2005.
41. X. Tang, D. Holland, A.M. Dale, L. Younes and M.I. Miller Shape abnormalities of subcortical and ventricular structures in mild cognitive impairment and Alzheimer's disease: detecting, quantifying, and predicting. *Human brain mapping*, 35 (8), 3701–3725, 2014.
42. D. Thompson. *On Growth and Form*. Dover reprint of 1942, 2nd ed. (1st ed., 1917), 1992.
43. E. Trélat. *Contrôle optimal: théorie et applications*. Vuibert, Collection *Mathématiques Concrètes*, 2nd ed., 2008.
44. A. Trouvé. Action de groupe de dimension infinie et reconnaissance de formes. *C R Acad Sci Paris Sér I Math*, 321(8):1031–1034, 1995.
45. A. Trouvé and F.-X. Vialard. Shape Splines and Stochastic Shape Evolutions: A Second Order Point of View. *Quarterly of Applied Mathematics*, page 26, 2010.
46. A. Trouvé and L. Younes. Local Geometry of Deformable Templates. *SIAM Journal on Mathematical Analysis*, 37(1):17, Nov. 2005.
47. M. Vaillant and J. Glaunes. Surface matching via currents. *Proceedings of Information Processing in Medical Imaging (IPMI), Lecture Notes in Computer Science*, 3565(381–392), 2006.
48. F.-X. Vialard, L. Risser, D. Rueckert, and C. Cotter. 3d image registration via geodesic shooting using and efficient adjoint calculation. *Journal International Journal of Computer Vision*, 97(2):229–241, April 2012.
49. T. Vercauteren, X. Pennec, A. Perchant, and N. Ayache. Diffeomorphic demons: Effcient non-parametric image registration. *NeuroImage*. 45(1):S61–S72, 2009.
50. L. Younes. *Shapes and Diffeomorphisms*, volume 171. Series: Applied Mathematical Sciences, 2010.

3D Normal Coordinate Systems for Cortical Areas

J. Tilak Ratnanather

Center for Imaging Science & Institute for Computational Medicine,
Department of Biomedical Engineering,
Johns Hopkins University, Baltimore MD 21218, USA
tilak@cis.jhu.edu

Sylvain Arguillère

CNRS, Université Claude Bernard Lyon 1 and CNRS UMR 5208,
Institut Camille Jordan, Lyon, France
sarguillere@gmail.com

Kwame S. Kutten

Center for Imaging Science, Johns Hopkins University,
Baltimore MD 21218, USA
kkutten1@jhmi.edu

Peter Hubka

Institute of AudioNeuroTechnology & Department of Experimental Otology,
Hannover Medical School, Hannover, Germany
hubka.peter@mh-hannover.de

Andrej Kral

Institute of AudioNeuroTechnology & Department of Experimental Otology,
Hannover Medical School, Hannover, Germany
kral.andrej@mh-hannover.de

Laurent Younes

*Center for Imaging Science & Institute for Computational Medicine,
Department of Applied Mathematics and Statistics,
Johns Hopkins University, Baltimore MD 21218, USA
younes@cis.jhu.edu*

A surface-based diffeomorphic algorithm to generate 3D coordinate grids in the cortical ribbon is described. In the grid, normal coordinate lines are generated by the diffeomorphic evolution from the gray/white (inner) surface to the gray/csf (outer) surface. Specifically, the cortical ribbon is described by two triangulated surfaces with open boundaries. Conceptually, the inner surface sits on top of the white matter structure and the outer on top of the gray matter. It is assumed that the cortical ribbon consists of cortical columns which are orthogonal to the white matter surface. This might be viewed as a consequence of the development of the columns in the embryo. It is also assumed that the columns are orthogonal to the outer surface so that the resultant vector field is orthogonal to the evolving surface. Then the distance of the normal lines from the vector field such that the inner surface evolves diffeomorphically towards the outer one can be construed as a measure of thickness. Applications are described for the auditory cortices in human adults and cats with normal hearing or hearing loss. The approach offers great potential for cortical morphometry.

1. Introduction

A conspicuous feature of the mammalian brain is the folded cortex, i.e. cortical ribbon, which maximises surface area within a confined space with the folding varying in degree from large mammals to small ones [1]. The cortex consists of neural tissue called gray matter, containing mostly neuronal cell bodies and unmyelinated fibers. The cortex can be divided into several regions or areas. For example, the human cortex has about 180 cortical regions [2] that serve different functional roles [3], vary in size and are connected via white matter containing axonal, usually myelinated, fibers. At the microscopic level, each cortical region is composed of fundamental units called cortical columns [4] that traverse vertically from the white matter to the surface just below the pia matter. The cortical column consists of six layers that are stacked horizontally on top of each other [5]. The orthogonality of cortical columns stems from the fact that in the embryonic stage, columns are formed by neurons migrating through the initially flat cortical plate which in turns folds as the brain expands [6].

Here cortical morphometry is more than volumetric analysis. The surface area of the interface with the white matter and thickness combine to yield the volume of the cortical cortex. Yet surface area and thickness can be differentially affected [7] in development and disease due to the influence of genes [3] and environment [8]. Expansion in surface area may be associated with increase in number of columns while increasing thickness may be associated with changes in the columnar microcircuitry [9].

Morphometric measures such as surface area, thickness, volume and curvature warrant a precise 3D coordinate system that reflects the columnar and laminar structure of the cortical area. Therefore it is necessary to construct a coordinate system that traverses normally between the inner and outer boundaries of the cortical ribbon. In other words, a continuous deformation of the inner surface onto the outer surface, using large deformations to accommodate the highly folded cortical ribbon as shown in the left panel of Fig. 1. Such a mapping can be developed in the Large Deformation Diffeomorphic Metric Mapping (LDDMM) framework [10] but with imposed normal constraints to enable the inner surface evolve to the outer surface. This follows but contrasts with orthogonal or curvilinear coordinate image-based systems proposed in recent years [11, 12, 13, 14].

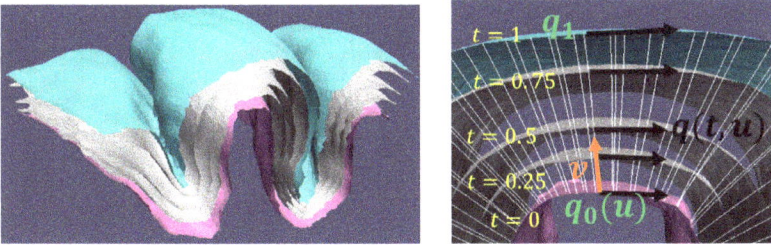

Fig. 1. Left: A cortical ribbon from a cat brain with a continuous deformation of the inner (pink) surface onto the outer (blue) surface described by the gray surfaces. Right: A 3D coordinate system for the cortical ribbon between the parametrized inner and outer surfaces (q_0 and q_1) is represented by the mapping $\Phi : U \times [0, 1] \to \mathbb{R}^3$ where the coordinates (u, t) of the inner surface are mapped to those of the evolving surface $q(t, u)$ indicated by three intermediate surfaces at $t = 0.25, 0.5$ and 0.75 and the vector field v that is orthogonal to the evolving surface.

2. Methods

2.1. *Theory*

The cortical region is endowed with normal coordinates defined by the apex and base of the cortical columns which in turn form natural coordinates for, i.e. parametrize, the inner and outer surfaces. Thus it is natural to view the columns tracing a diffeomorphism between two coordinate systems as illustrated in the right panel of Fig. 1. Consider two parametrized non-intersecting open surfaces: $q_0 : U \to \mathbb{R}^3$ (the inner surface) and $q_1 : U \to \mathbb{R}^3$ (the outer surface) where U is an open subset of $\mathbb{R}^2$ and q denotes a smooth embedding. The goal is to find a time-dependent surface parametrization $q(t) : U \to \mathbb{R}^3$ such that (i) $q(0) = q_0$ and $q(1)$ is a reparametrization of the surface associated with q_1; (ii) $q(t)$ is obtained from $q(0)$ via a time-dependent diffeomorphism, so that $q(t,x) = \varphi(t, q_0(x))$ for some $\varphi(t, \cdot)$; (iii) the associated Eulerian velocity $\partial_t q(t,x)$ is at all times perpendicular to the evolving surface.

Let V be a reproducing kernel Hilbert space (RKHS) of vector fields $v : \mathbb{R}^3 \to \mathbb{R}^3$, that is assumed to be, for some $p \geq 2$, continuously included in $C_0^p(\mathbb{R}^3, \mathbb{R}^3)$, the space of C^p vector fields with all partial derivatives up to order p tending to 0 at infinity, equipped with the usual supremum norm over all $x \in \mathbb{R}^d$ and all derivatives of order p or less. Let $v \mapsto \|v\|_V$ denote the Hilbert norm on V. Consider the shape space $\mathcal{M}$ of all C^1 embeddings $q : U \to \mathbb{R}^3$, and assume that $q_0, q_1 \in \mathcal{M}$. For $q \in \mathcal{M}$, let $S_q = q(U)$, which is, by assumption a submanifold of $\mathbb{R}^3$.

For $q \in \mathcal{M}$, let $V_q = \{v \in V : Dq(x)^t v(q(x)) = 0, x \in U\}$ denote the space of all vector fields in V that are perpendicular to the evolving surface S_q. Also, define, for $v \in V$, the surface-dependent norm

$$\|v\|_q^2 = \|v\|_V^2 + \lambda \int_{S_q} |Dv|^2 d\sigma_{S_q},$$

where Dv is the differential of v, $|Dv|^2$ is the squared Frobenius norm of Dv (the sum of its squared entries) and σ_{S_q} is the volume measure of S_q. Here $\|v\|_q$ is a hybrid norm [15] where the V norm, which is the standard norm in LDDMM, ensures that the transformation is diffeomorphic and penalizes non-local collisions within surfaces and the second term restricts local deformations on the surfaces. Combining the two terms permits larger penalties for local perturbations and milder ones for non-local collisions, which is well adapted for highly folded surfaces such as the ones considered here. Finally, assume that the chosen "data attachment" term is represented as a function $q \mapsto D(S_q, S_{q_1})$, differentiable in q. Examples of such terms are provided by including the space of surfaces in the dual of some suitable RKHS, leading to representations as measures [16], currents [17] or varifolds

[18]. Specifically, the latter is used with

$$D(S, S') = \int_S \int_{S'} \chi(p, p')(1 + (N_S(p)^t N_{S'}(p'))^2) \, d\sigma_S(p) \, d\sigma_{S'}(p')$$

where N_S and $N_{S'}$ are the unit normals to S and S', respectively and χ is a positive kernel. It is important that these data attachment terms depend only on the manifolds, S, S' and not on their parametrization, to ensure a reparametrization-invariant registration method. This is true for D because it is defined in terms of intrinsic properties of the surfaces embedded in $\mathbb{R}^3$.

With this notation, the following optimal control problem is solved. Minimize

$$F(q(\cdot), v(\cdot)) = \int_0^1 \|v(t)\|_{q(t)}^2 \, dt + D(S_{q(1)}, q_1),$$

subject to $q(0) = q_0$, $v(t) \in V_{q(t)}$ and $\partial_t q(t) = v(t, q(t))$. If $v(\cdot)$ is a solution, its flow φ, defined by $\partial_t \varphi(t) = v(t, \varphi(t))$, is a diffeomorphism such that $q(1) = \varphi(1, q_0)$. Computationally, the constraint $v \in V_q$ is enforced with an augmented Lagrangian [19] which has been used in constrained LDDMM problems [20].

The solution yields a 3D coordinate system for the space between q_0 and q_1 (cf. Fig. 1):

$$\Phi: \quad U \times [0, 1] \to \mathbb{R}^3$$
$$(u, t) \mapsto q(t, u).$$

(Recall that $u \in U$ is a 2D parameter.) The surfaces $t = $ constant can be interpreted as "sheets" between the inner and outer surfaces, and the transversal or normal lines $u = $ constant as "columnar lines". By no means, does this geometric construction actually represent the laminar and columnar properties of the cortical region. However the lengths of the normal lines can be used as measures of the thickness of the cortical region with respect to $q_0(u)$.

2.2. Data

Three datasets were used as to illustrate the theory. In the following, the inner and outer surfaces were obtained from triangulations of segmented cortices [21].

2.2.1. *Motor cortex*

Two MRI scans of a single subject with resolution $0.5 \times 0.5 \times 0.5$ mm/voxel taken a week apart were obtained from a 7T scanner (Philips Healthcare, The Netherlands). A subvolume of the primary motor cortex (M1) was manually segmented [22] from which the anterior portion was used for the analysis.

2.2.2. *Primary and higher order auditory cortices in cats*

MRI scans of two cats (one with hearing loss) with resolution $0.176 \times 0.176 \times 0.411$ mm/voxel were obtained from a 7T scanner (Bruker BioSpin, Germany). A subvolume encompassing the primary and higher order auditory cortices was segmented manually [23].

2.2.3. *Heschl's gyrus and Planum temporale in adults*

MRI scans of 10 adults (5 with profound hearing loss who consistently used Listening and Spoken Language with hearing aids after early diagnosis and intervention in infancy) were obtained from a 1.5T scanner (Phillips Healthcare, The Netherlands). Subvolumes encompassing Heschl's gyrus and Planum temporale were parcellated and segmented [24] with resolution $1 \times 1 \times 1$ mm/voxel.

3. Results

Figure 2 shows the normal lines for the M1 cortex in a single subject scanned at two time points a week apart. The distribution of the lengths of the normal lines from the two scans (not shown) are virtually similar. So the method is robust to perturbations such as scan times, segmentations, triangulation and delineation.

Figure 3 shows the results for one normal hearing cat and one congenitally deaf cat including zoomed views of one gyral crown and one sulcal fundus.

Since the lengths of the normal lines provide a notion of thickness, it is constructive to compare with those computed by widely-used brain mapping software, specifically FreeSurfer [25]. Given inner and outer surfaces S^I and S^O, the distribution of distances with respect to all vertices (denoted by r subscripted by the vertex index) of S^I computed by FreeSurfer

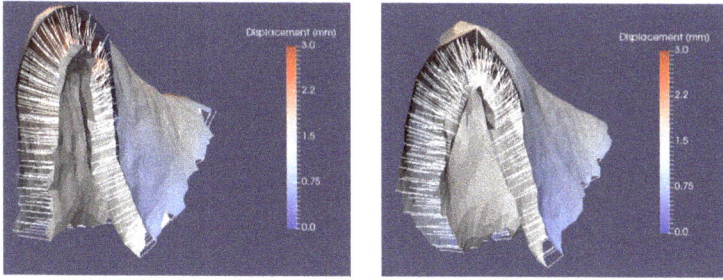

Fig. 2. Normal lines for the motor cortex in a single subject scanned one week apart. The color bar shows the variation in the lengths of the lines i.e a distribution of thickness.

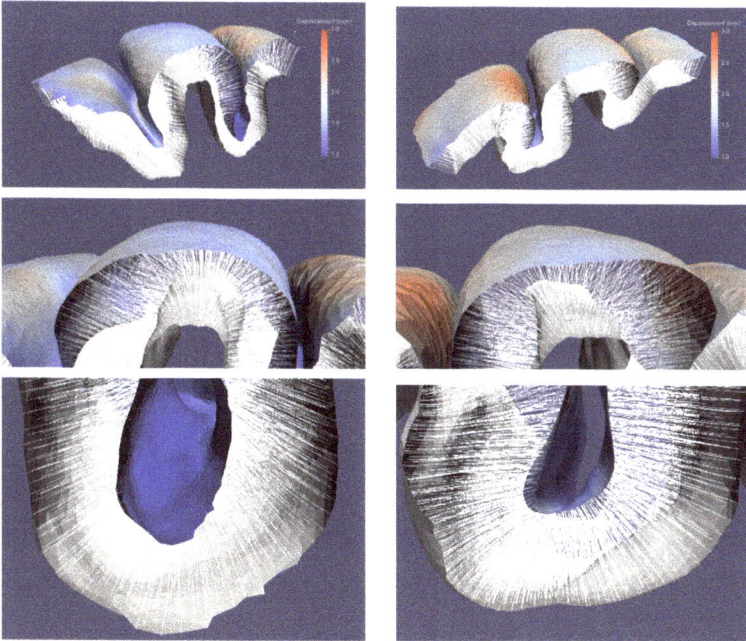

Fig. 3. Normal lines for primary and secondary auditory cortices in a normal hearing cat (left) and a congenitally deaf cat (right).

is

$$\left\{ \frac{\rho(r_i, f(r_i)) + \rho(f(r_i), g(f(r_i)))}{2} \forall r_i \in S^I \right\}$$

where $f(r_i) = \arg\min_{r_l \in S^O} |r_i - r_l|$, $g(f(r_i)) = \arg\min_{r_l \in S^I} |f(r_i) - r_l|$ and $\rho(r_a, r_b) = ||r_a - r_b||$ is the Euclidean distance. Figure 4 shows that FreeSurfer underes-

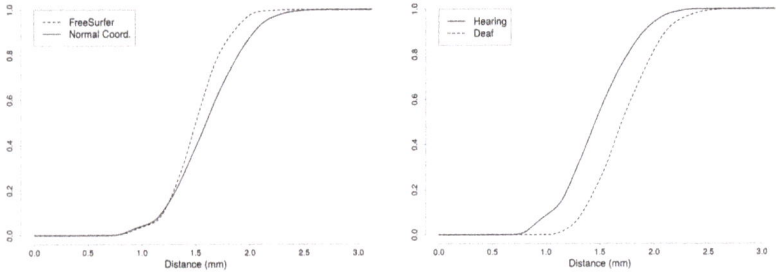

Fig. 4.   Left: FreeSurfer distances underestimates those computed from the normal lines. Right: Distances of normal lines in the congenitally deaf cat are bigger than those in the normal hearing cat.

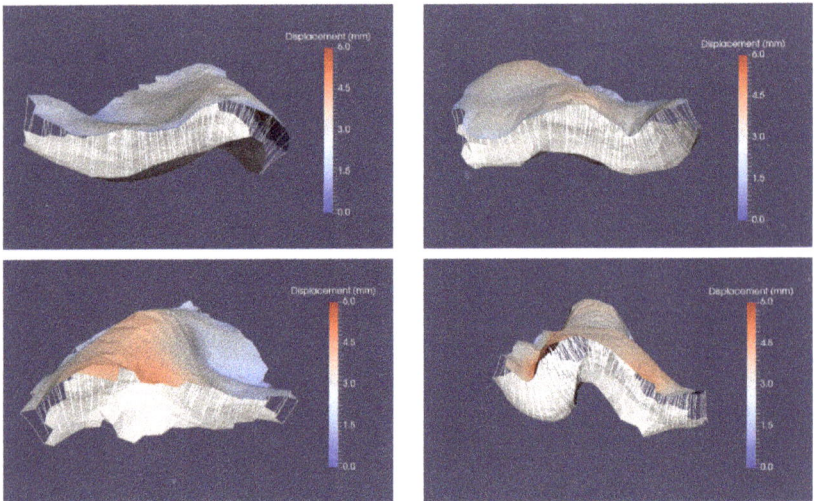

Fig. 5.   Left and right planum temporale in a subject with normal hearing (top) and hearing loss (bottom).

timates those computed by the present method and that the distribution of distances of the normal lines computed in Fig. 3 for the deaf cat is different from that of the hearing cat.

Figure 5 shows normal lines for the left and right planum tempo-rale in two adults - one with normal hearing and one with hearing loss. Figure 6 shows the distribution of distances; also shown are pooled distri-butions. Table 1 shows the performance of typical computations for different structures.

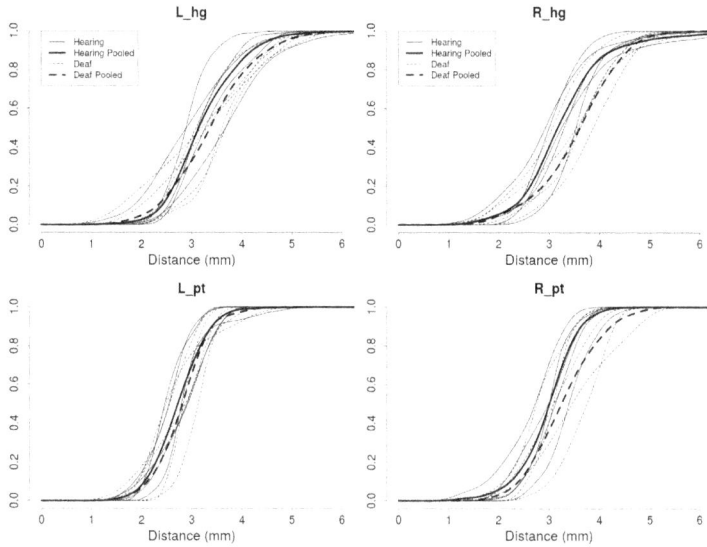

Fig. 6. Distances in the left and right Heschl's gyrus (top) and Planum temporale (bottom) in 10 adults (5 with profound hearing loss).

Table 1. Performance of typical computations

Cortex	Inner		Outer		Iterations	Runtime (hours)
	Vertices	Faces	Vertices	Faces		
M1	239	417	375	670	2417	1.45
Cat	4178	8011	4401	8396	606	24.53
PT	199	343	215	373	2343	1.17

4. Discussion

A columnar like coordinate system for cortical regions has been developed and applied in several cortices. The coordinates consists of the manifold evolving from the inner one to the outer one over $t \in [0, 1]$. But it will be necessary to use histological data to ascertain if the evolving manifold (surface) follows the laminar properties of the cortical layers [12, 11, 26, 27].

The approach adopted here exploits the geometric properties to generate the columnar like normal lines particularly in highly curved areas such as the gyral crowns and sulcal fundi which has been problematic in image-based approaches [11, 12, 13, 14] that also enforce orthogonality and equivolumetric laminar properties. In contrast, the triangulated inner surface forms a natural chart that is able to accommodate deeply buried sulci. For whole brain data, it may be easier to derive thickness from these image-based approaches but locally geometric based approaches such as the one described here may yield more reliable values. Even so, the method is computationally intensive in part due to using augmented Lagrangian to solve the optimal control problem [20]. It is not surprising that the run times are affected by number of faces and curvature. Speeding up computations could be achieved via a GPU or memory-efficient implementation of automatic differentiation libraries [28, 29, 30]. Thus this approach may be more suitable for cortical morphometry of regions of interest. So it would be interesting to evaluate the robustness of thickness computation with respect to different partitions of a larger area. Also an inexact matching method is used which means that the deformed inner surface does not exactly align with the target outer surface. There is therefore a small error in computing thickness due to the misalignment which could be corrected by adjusting the normal lines at their crossing points with the outer surface. Finally, normal lines from open boundaries can be distorted due to the irregular boundaries caused by the triangulation. Strategies to address this include expanding outside cortical areas or imposing additional normal constraints.

It is clear that cortical areas have distributions of thickness due to curvature with sulci and gyri having smaller and larger distances respectively. Such distributions are amenable to statistical analysis not too dissimilar to previous work on labeled cortical depth maps [31]. It is not surprising given that distances generated by FreeSurfer underestimates those generated here.

With respect to the primary and secondary auditory areas, the deaf cat is thought to be thicker than the normal hearing cat [32]. This is in contrast to the recent measurements of thickness based on histological data of gyral crowns [33] which excludes the folds shown here. Also, differences in the myelination of axonal fibers in the neighborhood of the gray/white matter interface could influence results [34].

There are limited opportunities for studying adults with hearing loss who have been using listening and spoken language (LSL) via hearing aids since infancy [35]. Such people are difficult to recruit primarily because many now have cochlear implants. Yet they are likely to have normal (or near-normal) P1 and N1 latencies (from cortical auditory evoked potential data) which correlate with neural activity associated with primary and secondary auditory cortices i.e. Heschl's gyrus and Planum temporale [36, 37, 38, 39]. The pooled distributions suggest little differences in the left side which is associated with temporal processing [40] and some differences on the right which is associated with spectral processing [40]. The former may be attributed to auditory training used in LSL after early detection and intervention as infants while the latter may be attributed to high frequency hearing loss. By comparison, thicker visual cortical areas have been observed in people blinded since infancy [41].

Future work will be focused initially on developing optimal sampling to build realistic 3D coordinates in cortical areas. In turn, these will be used for functional and stereological analysis particularly to assess the contributions from the laminar cortical layers to the overall properties of the cortical area.

Acknowledgments

Funding from the National Institutes of Health (P41-EB015909, R01-DC016784), Federal Ministry of Education and Research (BMBF) of Germany (01GQ1703), National Organization for Hearing Research for the scans of adults with hearing loss, Institute of Mathematical Sciences at the National University of Singapore, the Isaac Newton Institute for Mathematical Sciences at the University of Cambridge (EP/K032208/1) during the Growth Form and Self-organisation programme are gratefully acknowledged.

References

1. W. Welker. Why does cerebral cortex fissure and fold? In eds. E. G. Jones and A. Peters, *Cerebral Cortex: Comparative Structure and Evolution of Cerebral Cortex, Part II*, pp. 3–136. Springer (1990).
2. M. F. Glasser, T. S. Coalson, E. C. Robinson, C. D. Hacker, J. Harwell, E. Yacoub, K. Ugurbil, J. Andersson, C. F. Beckmann, M. Jenkinson, S. M. Smith, and D. C. Van Essen, A multi-modal parcellation of human cerebral cortex, *Nature*. **536**, 171–178 (2016).
3. P. Rakic, Specification of cerebral cortical areas, *Science*. **241**, 170 (1988).
4. V. B. Mountcastle, Modality and topographic properties of single neurons of cat's somatic sensory cortex, *J. Neurophysiol.* **20**, 408–434 (1957).
5. S. R. Cajal, *Comparative study of the sensory areas of the human cortex.* Clark University (1899).
6. D. H. Geschwind and P. Rakic, Cortical evolution: Judge the brain by its cover, *Neuron*. **80**, 633–647 (2013).
7. A. M. Winkler, D. N. Greve, K. J. Bjuland, T. E. Nichols, M. R. Sabuncu, A. K. Håberg, J. Skranes, and L. M. Rimol, Joint analysis of cortical area and thickness as a replacement for the analysis of the volume of the cerebral cortex, *Cerebral Cortex*. **28**, 738–749 (2018).
8. P. Rakic, Evolution of the neocortex: A perspective from developmental biology, *Nature Reviews Neuroscience*. **10**, 724–735 (2009).
9. K. Wagstyl and J. P. Lerch. Cortical thickness. In *Neuromethods*, pp. 35–49. Springer New York (2018).

10. M. F. Beg, M. I. Miller, A. Trouvé, and L. Younes, Computing large deformation metric mappings via geodesic flows of diffeomorphisms, *Int. J. Comp. Vision.* **61**, 139–157 (2005).

11. Y. Leprince, F. Poupon, T. Delzescaux, D. Hasboun, C. Poupon, and D. Rivière. Combined Laplacian-equivolumic model for studying cortical lamination with ultra high field MRI (7 T). In *Proc. IEEE 12th Int. Symp. Biomedical Imaging (ISBI)*, pp. 580–583, IEEE (2015).

12. M. D. Waehnert, J. Dinse, M. Weiss, M. N. Streicher, P. Waehnert, S. Geyer, R. Turner, and P.-L. Bazin, Anatomically motivated modeling of cortical laminae, *NeuroImage.* **93**, 210–220 (2014).

13. S. E. Jones, B. R. Buchbinder, and I. Aharon, Three-dimensional mapping of cortical thickness using Laplace's equation, *Human Brain Mapping.* **11**, 12–32 (2000).

14. S. R. Das, B. B. Avants, M. Grossman, and J. C. Gee, Registration based cortical thickness measurement, *NeuroImage.* **45**, 867–879 (2009).

15. L. Younes, Hybrid riemannian metrics for diffeomorphic shape registration, *Ann. Math. Sci. Appl.* **3**, 189–210 (2018).

16. J. Glaunès, A. Trouvé, and L. Younes. Diffeomorphic matching of distributions: A new approach for unlabelled point-sets and sub-manifolds matching. In *Proceedings of the 2004 IEEE Computer Society Conference on Computer Vision and Pattern Recognition.*, vol. 2, pp. 712–718, IEEE (2004).

17. M. Vaillant and J. Glaunès, Surface matching via currents, *Lect. Notes Comp. Sci.* **3565**, 381–392 (2005).

18. N. Charon and A. Trouvé, The varifold representation of nonoriented shapes for diffeomorphic registration, *SIAM J. Imaging Sciences.* **6**, 2547–2580 (2013).

19. J. Nocedal and S. J. Wright, *Numerical Optimization.* Springer (2006).

20. S. Arguillère, E. Trélat, A. Trouvé, and L. Younes, Shape deformation analysis from the optimal control viewpoint, *Journal de Mathématiques Pures et Appliquées.* **104**, 139–178 (2015).

21. S.-W. Cheng, T. K. Dey, and J. Shewchuk, *Delaunay Mesh Generation.* Chapman & Hall/CRC (2012).

22. J. K. Mai, M. Majtanik, and G. Paxinos, *Atlas of the Human Brain.* Academic Press (2015).

23. F. Reinoso-Suarez, *Topographischer Hirnatlas der Katze (Fur Experimentalphysiologische Untersuchungen). Herausgegeben von Merck AG.* Herausgegeben von E. merck AG, Darmstadt, Germany (1961).

24. J. T. Ratnanather, P. E. Barta, N. A. Honeycutt, N. Lee, H. M. Morris, A. C. Dziorny, M. K. Hurdal, G. D. Pearlson, and M. I. Miller, Dynamic programming generation of boundaries of local coordinatized submanifolds in the neocortex: application to the planum temporale, *NeuroImage.* **20**, 359–377 (2003).

25. B. Fischl, Freesurfer, *NeuroImage.* **62** (2), 774–781 (2012).

26. J. M. Huntenburg, K. Wagstyl, C. J. Steele, T. Funck, R. A. I. Bethlehem, O. Foubet, B. Larrat, V. Borrell, and P.-L. Bazin, Laminar Python: Tools

for cortical depth-resolved analysis of high-resolution brain imaging data in Python, *Research Ideas and Outcomes.* **3** (2017).

27. S. T. Bok, *Histonomy of the Cerebral Cortex.* Elsevier Pub. Co. (1959).

28. B. Charlier, J. Feydy, and J. Glaunès. KeOps: Calcul rapide sur GPU dans les espaces à noyaux. In *Proceedings of Journées de Statistique de la SFdS*, Société Française de Statistique, Paris, France (2018). URL `https://toltex.u-ga.fr/users/RCqls/Workshop/jds2018/resumesLongs/subm309.pdf`.

29. D. Tward, M. Miller, and Alzheimer's Disease Neuroimaging Initiative. Unbiased diffeomorphic mapping of longitudinal data with simultaneous subject specific template estimation. In *Graphs in Biomedical Image Analysis, Computational Anatomy and Imaging Genetics*, pp. 125–136. Springer (2017).

30. L. Kühnel and S. Sommer. Computational anatomy in Theano. In *Graphs in Biomedical Image Analysis, Computational Anatomy and Imaging Genetics*, pp. 164–176. Springer (2017).

31. E. Ceyhan, M. Hosakere, T. Nishino, J. Alexopoulos, R. D. Todd, K. N. Botteron, M. I. Miller, and J. T. Ratnanather, Statistical analysis of cortical morphometrics using pooled distances based on labeled cortical distance maps, *J. Math. Imaging and Vision.* **40**, 20–35 (2010).

32. P. Barone, L. Lacassagne, and A. Kral, Reorganization of the connectivity of cortical field DZ in congenitally deaf cat, *PLoS ONE.* **8**, e60093 (2013).

33. C. Berger, D. Kühne, V. Scheper, and A. Kral, Congenital deafness affects deep layers in primary and secondary auditory cortex, *J. Comp. Neurol.* **525**, 3110–3125 (2017).

34. K. Emmorey, J. S. Allen, J. Bruss, N. Schenker, and H. Damasio, A morphometric analysis of auditory brain regions in congenitally deaf adults, *Proc. Nat. Acad. Sci.* **100**, 10049–10054 (2003).

35. O. A. Olulade, D. S. Koo, C. J. LaSasso, and G. F. Eden, Neuroanatomical profiles of deafness in the context of native language experience, *J. Neurosci.* **34**, 5613–5620 (2014).

36. A. Sharma, K. Martin, P. Roland, P. Bauer, M. H. Sweeney, P. Gilley, and M. Dorman, P1 latency as a biomarker for central auditory development in children with hearing impairment, *J. Am. Acad. Audiol.* **16**, 564–573 (2005).

37. A. Kral and A. Sharma, Developmental neuroplasticity after cochlear implantation, *Trends in Neurosciences.* **35**, 111–122 (2012).

38. A. Sharma, J. Campbell, and G. Cardon, Developmental and cross-modal plasticity in deafness: Evidence from the P1 and N1 event related potentials in cochlear implanted children, *Int. J. Psychophysiol.* **95**, 135–144 (2015).

39. F. Liem, T. Zaehle, A. Burkhard, L. Jäncke, and M. Meyer, Cortical thickness of supratemporal plane predicts auditory N1 amplitude, *NeuroReport.* **23**, 1026–1030 (2012).

40. D. Marie, S. Maingault, F. Crivello, B. Mazoyer, and N. Tzourio-Mazoyer, Surface-based morphometry of cortical thickness and surface area associated with Heschl's gyri duplications in 430 healthy volunteers, *Frontiers in Human Neuroscience.* **10** (2016).

41. J. Jiang, W. Zhu, F. Shi, Y. Liu, J. Li, W. Qin, K. Li, C. Yu, and T. Jiang, Thick visual cortex in the early blind, *J. Neurosci.* **29**, 2205–2211 (2009).

Heat Kernel Smoothing in Irregular Domains

Moo K. Chung* and Yanli Wang†

*Department of Biostatistics and Medical Informatics,
University of Wisconsin, Madison, WI 53706, USA
mkchung@wisc.edu

†Institute of Applied Physics and Computational Mathematics,
P. O. Box 8009, Beijing 100088, P. R. China
wangyanliwyl@gmail.com

We review the heat kernel smoothing techniques for denoising and re-gressing data in irregularly shaped domains embedded in Euclidean spaces. This is a problem often encountered in functional data analysis and medical imaging. In this chapter, we present a unified mathematical framework based on the eigenfunctions of the Laplace-Beltrami opera-tors defined on irregular domains. Numerical implementation issues will be addressed as well. Various examples will be presented. We also present a few new theoretical results on the properties of heat kernel smoothing.

Contents

1. Introduction

For *irregular domains* often encountered in images, boundary shapes are often very complex. This causes the geometric shape of the boundary to strongly affect the shape of Gaussian kernels. In such irregular domains, the use of Gaussian kernels may not be appropriate. So there is need to incorporate the shape of the boundary into the shape of kernels. The traditional methods include domain embedding methods which embed the domain of interest within a 2D rectangle or 3D box, where the usual sine and cosine basis are known [8]. Such methods still introduce the ringing artifacts (Gibbs phenomenon) along the boundary [12].

Heat kernel has been popular in shape modeling in recent years. Heat kernel is often used as a natural generalization of the Gaussian kernel. [3, 4] used the truncated Gaussian kernel in locally approximating the heat kernel in manifold learning. [45] used the heat kernel as a multiscale shape feature for surface meshes. [6] used the heat kernel signature (HKS), which is the trace of heat kernel, as an isometry-invariant multi-scale shape descriptor in computer vision. [31] computed the heat kernel on graphs using graph Laplacian for face representation. Also there have been significant developments in kernel methods in machine learning [39, 34, 41, 44, 51]. Most kernel methods in machine learning deal with the linear combination of kernels as a solution to penalized regressions. In most applications, the discrete versions of heat kernel, which in turn uses the eigenvector of graph Laplacian, are often used [42]. The connection between the eigenfunctions of continuous and discrete Laplacians has been well established by several studies [23, 46].

Heat kernel smoothing was introduced in [17, 16] to filter out noisy cortical thickness defined on brain surface mesh vertices obtained from magnetic resonance images [17, 16] approximates the heat kernel locally by iteratively applying Gaussian kernel with smaller bandwidth. For recent spectral formulation to heat kernel smoothing [40, 15] constructs the heat kernel analytically using the eigenfunctions of the Laplace-Beltrami (LB) operator, avoiding the need for the linear approximation using Gaussian kernel.

We will start by reviewing the basic spectral geometry related to heat kernel.

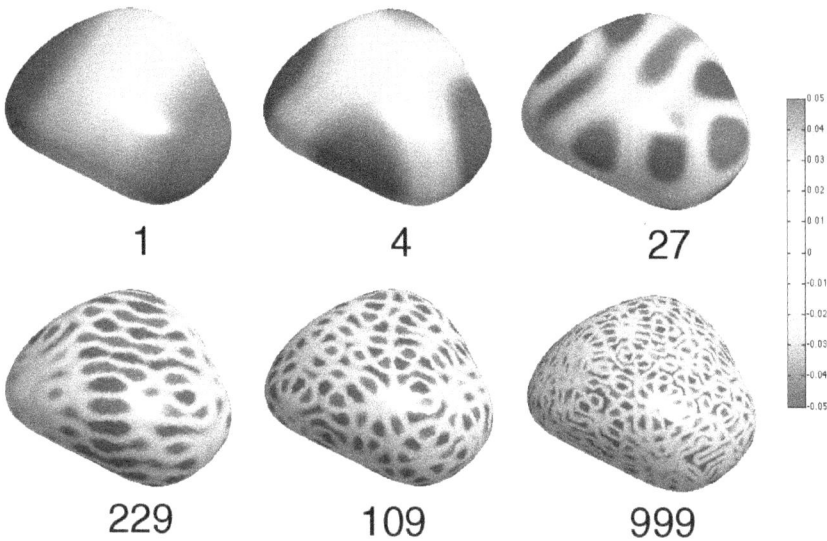

Fig. 1. Six representative eigenfunctions on human amygdala surface. The number j represents eigenfunction ψ_j. The eigenfunctions of the Laplace-Beltrami operators are computed using the cotan formulation.

2. Laplace-Beltrami eigenfunctions

Let Δ be the Laplace-Beltrami (LB) operator in a reasonably smooth manifold $\mathcal{M}$ in $\mathbb{R}^n$. The LB-operator associated with the Riemannian metric $g = (g_{ij})$ is then given by [29, 33]

$$\Delta = \frac{1}{|g|^{1/2}} \sum_{i,j} \frac{\partial}{\partial x_i} \left(|g|^{1/2} g^{ij} \frac{\partial}{\partial x_j} \right). \tag{2.1}$$

Solving the eigenvalue equation

$$\Delta \psi_j(p) = \lambda \psi_j(p), \; p \in \mathcal{M}, \tag{2.2}$$

we obtain ordered eigenvalues

$$0 = \lambda_0 < \lambda_1 \leq \lambda_2 \leq \cdots$$

and corresponding eigenfunctions $\psi_0, \psi_1, \psi_2, \ldots$. The first eigenvalue and eigenfunction are trivially given as $\lambda_0 = 0$ and $\psi_0 = 1/\sqrt{\mu(\mathcal{M})}$, where $\mu(\mathcal{M})$ is the volume of $\mathcal{M}$. It is possible to have multiple eigenfunctions corresponding to the same eigenvalue. The multiplicity usually happens when $\mathcal{M}$ has symmetry. However, if there is no symmetry, all the eigenvalues

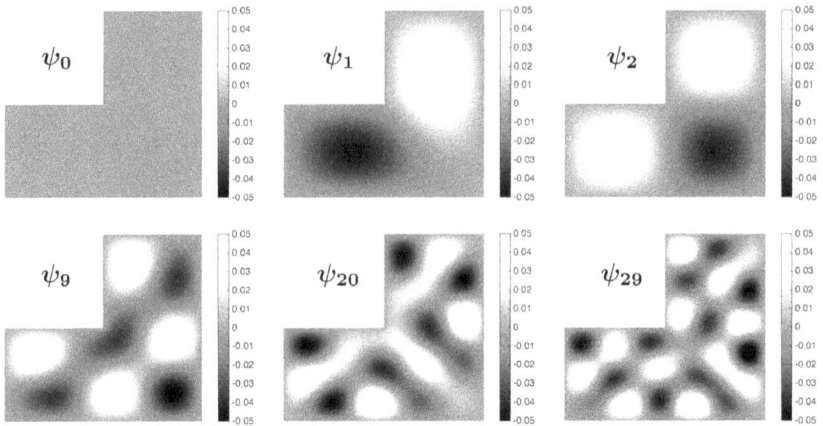

Fig. 2.   Eigenfunctions of the Laplacian in L-shaped domain.

are unique. The eigenfunctions ψ_j form an orthonormal basis in $L^2(\mathcal{M})$ [30, 38].

Other than $\psi_0 = 1/\sqrt{\mu(M)}$, there is no known close form expressions for eigenfunctions. For most shape modeling applications in medical imaging and computer vision, the underlying domains can be discretized using triangular or tetrahedral meshes; we can discretize the Laplace-Beltrami operator using the cotan formulation as the generalized eigenvalue problem[a] [18, 35, 40]:

$$\mathbf{C}\psi = \lambda \mathbf{A}\psi, \tag{2.3}$$

where $\mathbf{C}$ is the stiffness matrix, $\mathbf{A}$ is the mass matrix, and ψ is the unknown eigenfunction evaluated at mesh vertices. The details on the matrices $\mathbf{C}$ and $\mathbf{A}$ can be found in [11, 18, 35]. For tetrahedral meshes in $\mathbb{R}^3$, a similar cotan discretization is available [20, 47, 48]. Figure 1 displays few LB-eigenfunctions on the triangle meshes of human amygdala. Figure 2 displays the LB-eigenfunctions on the L-shaped domain.

2.1. *Graph Laplacian*

In computer vision and other sciences, the discrete version of Laplace-Beltrami operator is often used on graph data structures. Even if we do

---

[a]MATLAB code is available at http://brainimaging.waisman.wisc.edu/~chung/lb.

not have graphs, by subdividing regions into a collection of connected polygons, we can obtain graph Laplacian. Almost all the result of continuous counterpart is carried over to the discrete version. Here we explain the main difference in the graph Laplacian.

Let $G = \{V, E\}$ be a graph with vertex set V and edge set E. We will simply index the node set as $V = \{1, 2, \ldots, p\}$. If two nodes i and j form an edge, we denote it as $i \sim j$. Let $W = (w_{ij})$ be the edge weight. The adjacency matrix of G is often used as the edge weight. Various forms of graph Laplacian have been proposed [10] but the most often used standard form $L = (l_{ij})$ is given by

$$
l_{ij} = \begin{cases} -w_{ij}, & i \sim j \\ \sum_{i \neq j} w_{ij}, & i = j \\ 0, & \text{otherwise.} \end{cases} \tag{2.4}
$$

Even cotan discretization of LB-operator can be written in the form (2.4) [11, 18]. The graph Laplacian L can then be written in a matrix form

$$
L = D - W,
$$

where $D = (d_{ij})$ is the diagonal matrix with $d_{ii} = \sum_{j=1}^{n} w_{ij}$. For this chapter, we will simply use the adjacency matrix so that the edge weights w_{ij} are either 0 or 1.

Unlike the continuous Laplace-Beltrami operators that may have possibly infinite number of eigenfunctions, we have up to p number of eigenvectors

$$
\psi_1, \psi_2, \ldots, \psi_p
$$

satisfying

$$
L\psi_j = \lambda_j \psi_j \tag{2.5}
$$

with

$$
0 = \lambda_1 < \lambda_2 \leq \cdots \leq \lambda_p.
$$

The eigenvectors are orthonormal, i.e., $\psi_i^{\mathsf{T}} \psi_j = \delta_{ij}$, the Kroneker's delta. The first eigenvector is trivially given as $\psi_1 = 1/\sqrt{p}$ with $1 = (1, 1, \ldots, 1)^{\mathsf{T}}$.

All other higher order eigenvalues and eigenvectors are unknown analytically and have to be computed numerically. Using the eigenvalues and eigenvectors, the graph Laplacian can be decomposed spectrally. From (2.5),

$$
L\Psi = \Psi\Lambda, \tag{2.6}
$$

where $\Psi = [\psi_1, \ldots, \psi_p]$ and Λ is the diagonal matrix with entries $\lambda_1, \ldots, \lambda_p$. Since Ψ is an orthogonal matrix,

$$\Psi\Psi^\mathsf{T} = \Psi^\mathsf{T}\Psi = \sum_{j=1}^{p} \psi_j\psi_j^\mathsf{T} = I_p,$$

the identify matrix of size p. Then (2.6) is written as

$$L = \Psi\Lambda\Psi^\mathsf{T} = \sum_{j=1}^{p} \lambda_j\psi_j\psi_j^\mathsf{T}.$$

This is the restatement of the singular value decomposition (SVD) for Laplacian.

2.2. Laplacian in positive definite symmetric matrices

Recently in relation to brain network analysis and in diffusion tensor imaging, positive definite symmetric (PDS) matrices have become fundamental object of interest. PDS matrices would be considered as an irregular domain since the usual Euclidean geometry does not apply. Let $\mathcal{P}_m \subset \mathbb{R}^{m(m+1)/2}$ be the space of positive definite symmetric matrices of size $m \times m$.

For $Y = (y_{ij}) \in \mathcal{P}_m$, let $dY = (dy_{ij})$. Following [32], we will put the following metric on $\mathcal{P}_m$:

$$(ds)^2 = \operatorname{tr}\left((Y^{-1}dY)^2\right).$$

Vectorize $n = m(m+1)/2$ unique entries of Y as $(x_1, x_2, \ldots, x_n)'$ and write ds^2 in the standard quadratic form as

$$(ds)^2 = \sum g_{ij}dx_i dx_j.$$

For $\mathcal{P}_m$, this can be more compactly written as follows. Define the matrix of differential operators ∂ as

$$\partial_Y = \left(\frac{1}{2}(1 + \delta_{ij})\frac{\partial}{\partial y_{ij}}\right),$$

where δ_{ij} is Kronecker's delta. With this operator, the LB-operator Δ in the local coordinates y_{ij} is given by [24, 36]

$$\Delta = \operatorname{tr}(Y\partial_Y)^2. \tag{2.7}$$

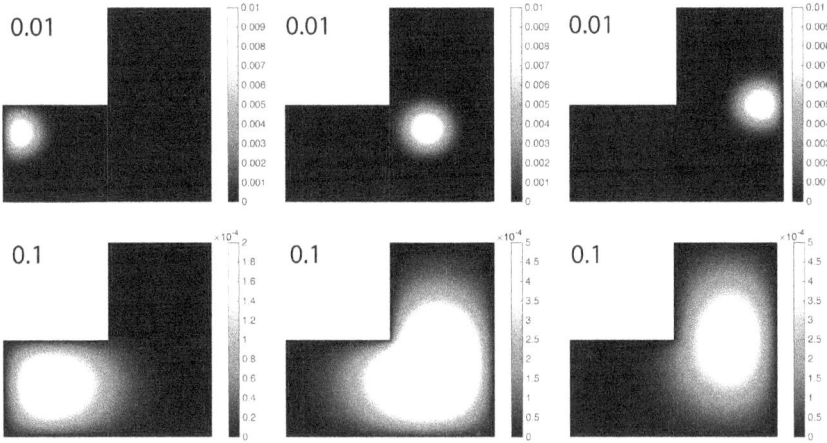

Fig. 3. Heat kernel in the L-shaped domain for two different bandwidths $\sigma = 0.01, 0.1$. We have used degree 70 expansions but the shape is almost identical if we use higher degree expansions.

Note that the Laplacian in the coordinates of the eigenvalues of y is more complicated from [28]. The eigenfunction of the Laplacian (2.7) is difficult to compute in practice and involves Zonal spherical functions [36, 37].

Example 2.1: Consider $\mathcal{P}_1 = \mathbb{R}^+$, the positive real line. Note that the Laplacian is parameterization invariant. Let $y = e^x$ to be the parameterization of $\mathcal{P}_1$. It maps $\mathbb{R}$ to $\mathbb{R}+$. Then $dy = y dx$ and with respect to the original coordinates y, we obtain (2.8).

$$\Delta = \left(\frac{d}{dx} \right)^2 = \left(y \frac{d}{dy} \right)^2 = y \frac{d}{dy} + y^2 \frac{d^2}{dy^2}. \tag{2.8}$$

Note Laplacian (2.8) in $\mathcal{P}_1$ differs from the usual Laplacian $\frac{d^2}{dy^2}$ for the whole real line. This additional algebraic complexity of the Laplacian makes the computation of eigenfunctions of even 1D case (2.8) complicated. In fact, we need to solve

$$y \frac{d}{dy} \psi_j(y) + y^2 \frac{d^2}{dy^2} \psi_j(y) = \lambda_j \psi_j(y). \tag{2.9}$$

In practice, it might be much easier to simply discretize the differential equation (2.9) and solve using the finite element method.

3. Heat kernel

The *heat kernel* $K_\sigma(p, q)$ is defined as

$$K_\sigma(p, q) = \sum_{j=0}^{\infty} e^{-\lambda_j \sigma} \psi_j(p) \psi_j(q), \tag{3.1}$$

where σ is the bandwidth of the kernel. The detailed mathematical exposition of heat kernel is given in [5] and [38]. Note that the heat kernel is the fundamental solution of an isotropic heat diffusion.

Symmetric kernel $G(p, q)$ defined on $\mathcal{M}$ is positive definite if

$$\sum_{i,j=1}^{n} G(p_i, q_j) c_i c_j > 0$$

for all choices of $p_i, q_j \in \mathcal{M}$ and nonzero $c_i, c_j \in \mathbb{R}$. Generalizing the definition, kernel $G(p, q)$ is *integrally positive definite* on $\mathcal{M}$ if

$$\int_{\mathcal{M}} G(p, q) f(p) f(q) \, d\mu(p) d\mu(q) > 0$$

for any $f \in L_1(\mathcal{M})$, the space of integrable functions. For a continuous kernel, these two definitions can be shown to be equivalent.

The heat kernel is a probability distribution, i.e.,

$$\int_{\mathcal{M}} K_\sigma(p, q) \, d\mu(p) = \int_{\mathcal{M}} K_\sigma(p, q) \, d\mu(q) = 1.$$

Thus, the discretized kernel matrix can be viewed as doubly-stochastic.

Figure 3 shows the heat kernel for bandwidth 0.01 and 1 for L-shaped domain. The kernel follows the shape of the irregular domain.

3.1. *Heat kernel on spheres*

On a two-sphere, the heat kernel is analytically given in terms of the spherical harmonics Y_{lm} [14]:

$$K_\sigma(p, q) = \sum_{l=0}^{\infty} \sum_{m=-l}^{l} e^{-l(l+1)\sigma} Y_{lm}(p) Y_{lm}(q). \tag{3.2}$$

On a three-sphere, the heat kernel is analytically given in terms of the hyperspherical harmonics Z_{nlm} [21, 27, 26, 25]:

$$K_\sigma(p, q) = \sum_{l=0}^{\infty} \sum_{m=-l}^{l} e^{-l(l+2)\sigma} Z_{lmn}(p) Z_{lmn}(q).$$

The hyperspherical harmonics $Z_{lmn}(p)$ with $p = (\beta, \theta, \phi)$ are given by

$$Z_{lmn}(\beta, \theta, \phi) = 2^{l+1/2}\sqrt{\frac{(n+1)\Gamma(n-l+1)}{\pi\Gamma(n+l+2)}}\Gamma(l+1)\sin^l\beta \, G_{n-l}^{l+1}(\cos\beta)Y_{lm}(\theta, \phi),$$

where $(\beta \in [0, \pi], \theta \in [0, \pi], \phi \in [0, 2\pi])$, G_{n-1}^{l+1} are the Gegenbauer (ultraspherical) polynomials, and Y_{lm} are the 3D spherical harmonics. The Gegenbauer polynomials can be expressed in terms of the Gaussian (ordinary) hypergeometric function:

$$G_\alpha^\lambda(x) = \frac{\Gamma(\alpha + 2\lambda)}{\alpha!\Gamma(2\lambda)}{}_2F_1(-\alpha, \alpha + 2\lambda; \lambda + \frac{1}{2}; \frac{1}{2}(1-x)).$$

The hyperpsherical harmonics form an orthonormal basis on the hypersphere:

$$\int_0^{2\pi}\int_0^\pi\int_0^\pi Z_{lmm}(\beta, \theta, \phi)Z_{l'm'n'}(\beta, \theta, \phi)\sin^2\beta \sin\theta d\beta d\theta d\phi = \delta_{nn'}\delta_{ll'}\delta_{mm'}.$$

3.2. Heat kernel on graphs

On a graph, the *discrete heat kernel* K_σ has a finite expansion and the simplicity in the algebraic representation makes it easier to manipulate. K_σ is a positive definite symmetric matrix of size $p \times p$ given by

$$K_\sigma = \sum_{j=1}^p e^{-\lambda_j\sigma}\psi_j\psi_j^\mathsf{T}, \tag{3.3}$$

where σ is called the bandwidth of the kernel. Alternately, we can write (3.3) as

$$K_\sigma = \Psi e^{-\sigma\Lambda}\Psi^\mathsf{T},$$

where $e^{-\sigma\Lambda}$ is the matrix logarithm of Λ. To see positive definiteness of the kernel, for any nonzero $x \in \mathbb{R}^p$, note

$$x^\mathsf{T}K_\sigma x = \sum_{j=1}^p e^{-\lambda_j\sigma}x^\mathsf{T}\psi_j\psi_j^\mathsf{T}x$$

$$= \sum_{j=1}^p e^{-\lambda_j\sigma}(\psi_j^\mathsf{T}x)^2 > 0.$$

When $\sigma = 0$, $K_0 = I_p$, identity matrix. When $\sigma = \infty$, by interchanging the sum and the limit, we obtain

$$K_\infty = \psi_1\psi_1^\mathsf{T} = \mathbf{1}\mathbf{1}^\mathsf{T}/p.$$

K_∞ is a degenerate case and the kernel is no longer positive definite. Other than these specific cases, the heat kernel is not analytically known in arbitrary graphs.

Heat kernel is doubly-stochastic [10] so that

$$K_\sigma \mathbf{1} = \mathbf{1}, \ \mathbf{1}^\mathsf{T} K_\sigma = \mathbf{1}^\mathsf{T}.$$

Thus, K_σ is a probability distribution along columns or rows.

Just like the continuous counterpart, the discrete heat kernel is also multiscale and has the scale-space property. Note

$$K_\sigma^2 = \sum_{i,j=1}^p e^{-(\lambda_i + \lambda_j)\sigma} \psi_i \psi_i^\mathsf{T} \psi_j \psi_j^\mathsf{T}$$
$$= \sum_{j=1}^p e^{-2\lambda_j \sigma} \psi_j \psi_j^\mathsf{T} = K_{2\sigma}.$$

We used the orthonormality of eigenvectors. Subsequently, we have

$$K_\sigma^n = K_{n\sigma}$$

for any integer $n \geq 0$.

4. Heat kernel smoothing

The concept of heat kernel smoothing was introduced in [17, 16] in the context of smoothing human cortical surface data. The original formulation used the tangent space approximation. The spectral version using LB-eigenfunction was later developed [40, 15]. *Heat kernel smoothing* of functional measurement $f(p)$ is then defined as

$$K_\sigma * f(p) = \int_\mathcal{M} K_\sigma(p, q) f(q) \, d\mu(q)$$
$$= \sum_{j=0}^\infty e^{-\lambda_j \sigma} f_j \psi_j(p),$$

where

$$f_j = \langle f, \psi_j \rangle = \int_\mathcal{M} f(p) \psi_j(p) \, d\mu(p)$$

are Fourier coefficients [17]. It is well known that heat kernel smoothing is the unique solution of an isotropic heat diffusion [38].

Theorem 4.1: *For an arbitrary self-adjoint differential operator Δ and*

$f \in L^2(\mathcal{M})$, *the unique solution of the Cauchy problem*

$$\frac{\partial g(p, \sigma)}{\partial \sigma} + \Delta g(p, \sigma) = 0, g(p, \sigma = 0) = f(p) \tag{4.1}$$

is given by

$$g(p, \sigma) = \sum_{j=0}^{\infty} e^{-\lambda_j \sigma} f_j \psi_j(p). \tag{4.2}$$

Proof: The statement is first given in [13] with heuristic proof. Here we provide a more rigorous proof. We first prove $g(p, \sigma) \in L^2(\mathcal{M})$. Since $g(p, \sigma)$ is the solution to (4.1), multiplying (4.1) with $g(p, \sigma)$ and integrating on $[0, T] \times \mathcal{M}$, we obtain

$$\int_0^T \int_{\mathcal{M}} \frac{\partial g(p, \sigma)}{\partial t} g(p, \sigma) \, d\mu(p) \, d\sigma + \int_0^T \int_{\mathcal{M}} \Delta g(p, \sigma) g(p, \sigma) \, d\mu(p) \, d\sigma = 0,$$

where $T > 0$ is total diffusion time. Hence, it holds that

$$\frac{1}{2} \int_{\mathcal{M}} \left(g(p, T)^2 - g(p, 0)^2 \right) \, d\mu(p) + \int_0^T \langle \Delta g, g \rangle \, d\sigma = 0.$$

Since Δ is self-adjoint,

$$\langle \Delta g, g \rangle \geq 0.$$

Thus,

$$\begin{aligned}
\|g(\cdot, T)\|_2^2 &= \int_{\mathcal{M}} g(p, T)^2 \, d\mu(p) \\
&= \int_{\mathcal{M}} g(p, 0)^2 \, d\mu(p) - 2 \int_0^T \langle \Delta g, g \rangle \, d\sigma \\
&\leq \int_{\mathcal{M}} g(p, 0)^2 \, d\mu(p) = \int_{\mathcal{M}} f(p)^2 \, d\mu(p) = \|f\|_2^2 < \infty,
\end{aligned}$$

where $\|\cdot\|_2$ is the L^2-norm.

Since eigenfunctions ψ_j form an orthonormal basis in $\mathcal{M}$, for each fixed σ, $g(p, \sigma)$ can be *uniquely* written as a Fourier series

$$g(p, \sigma) = \sum_{j=0}^{\infty} c_j(\sigma) \psi_j(p). \tag{4.3}$$

Then

$$\Delta g(p, \sigma) = \sum_{j=0}^{\infty} c_j(\sigma) \lambda_j \psi_j(p). \tag{4.4}$$

Substituting (4.3) and (4.4) into (4.1), we obtain

$$\frac{\partial c_j(\sigma)}{\partial \sigma} + \lambda_j c_j(\sigma \Delta a) = 0 \qquad (4.5)$$

for all j. The solution of equation (4.5) is given by $c_j(\sigma) = b_j e^{-\lambda_j \sigma}$. So we have a solution

$$g(p, \sigma) = \sum_{j=0}^{\infty} b_j e^{-\lambda_j \sigma} \psi_j(p).$$

At $\sigma = 0$, we have

$$g(p, 0) = \sum_{j=0}^{\infty} b_j \psi_j(p) = f(p).$$

The coefficients b_j must be given by the *unique* Fourier coefficients, i.e.,

$$b_j = \langle f, \psi_j \rangle = f_j.$$

$\square$

Since heat kernel smoothing is the solution of diffusion equation, it satisfies the scale-space property.

Theorem 4.2: *Denote the k-fold iterated kernel as*

$$K_\sigma^{(k)} = \underbrace{K_\sigma * \cdots * K_\sigma}_{k \text{ times}}.$$

Then we have

$$K_{k\sigma} * f = K_\sigma^{(k)} * f. \qquad (4.6)$$

Proof: $K_\sigma^{(2)} * f$ is equivalent to the diffusion of signal f after time 2σ. Hence we have

$$K_\sigma^{(2)} f = K_{2\sigma} * f.$$

Arguing inductively we see that the general statement holds. $\square$

Note an alternate proof can be obtained by noting that $K_\sigma^{(k)}$ is the density of the sum of k independent and identically distributed random variables in $\mathcal{M}$. Heat kernel with a large bandwidth is equivalent to the multiple applications of heat kernel smoothing with a smaller bandwidth. The property (4.6) was used to approximate heat kernel smoothing with

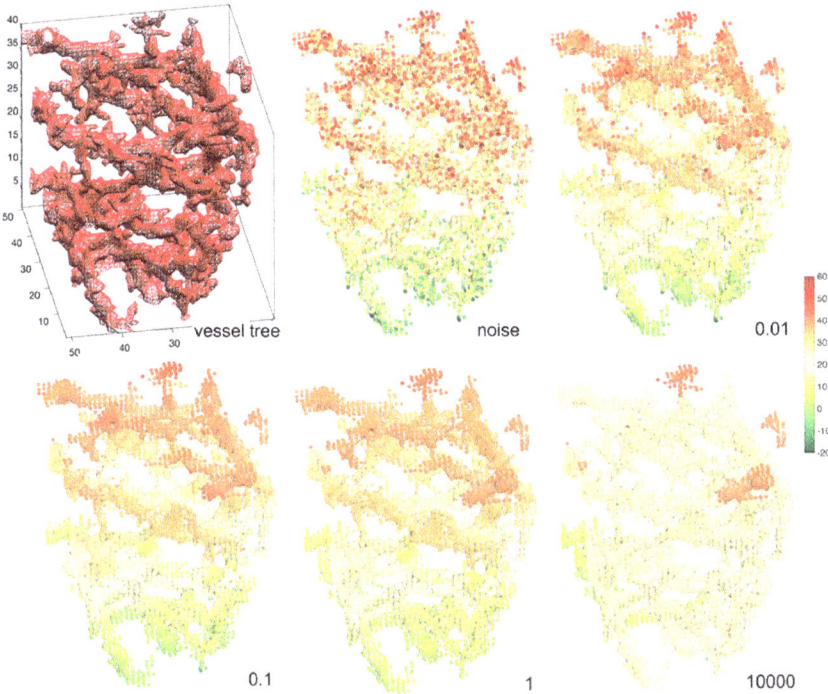

Fig. 4. Top left to right: 3D lung vessel tree. Gaussian noise is added to one of the coordinates. Heat kernel smoothing with bandwidth 0.01, 0.1, 1 and 10000.

multiple applications of Gaussian kernel smoothing with small bandwidth [17]. If we change the scale to $2\sigma = \tau^2$, (4.6) takes a slightly different form:

$$K_\tau^{(k)} * f = K_{\sqrt{k}\tau} * f.$$

4.1. *Asymptotics*

As $\sigma \to 0$, $K_\sigma(p, q)$ becomes the Dirac-delta function $\delta(p - q)$ so the heat kernel smoothing becomes unbiased as $\sigma \to 0$, i.e.,

$$\lim_{\sigma \to 0} K_\sigma * f(p) = f(p).$$

Theorem 4.3: *For $f \in L^2(\mathcal{M})$ with $\mu(\mathcal{M}) \leq \infty$, the heat kernel smoothing converges to the mean signal over $\mathcal{M}$ pointwisely*

$$\lim_{\sigma \to \infty} K_\sigma * f(p) = \frac{1}{\mu(\mathcal{M})} \int_{\mathcal{M}} f(p) \, d\mu(p),$$

for all $p \in \mathcal{M}$.

Proof: The statement is given in [17, 38]. Since $K_\sigma * f$ is bounded, we can interchange the limit and summation.

$$\lim_{\sigma \to \infty} K_\sigma * f = \lim_{\sigma \to \infty} \sum_{j=0}^{\infty} e^{-\lambda_j \sigma} f_j \psi_j(p)$$

$$= \sum_{j=0}^{\infty} \lim_{\sigma \to \infty} e^{-\lambda_j \sigma} f_j \psi_j(p)$$

$$= f_0 \psi_0(p)$$

$$= \frac{1}{\mu(\mathcal{M})} \int_{\mathcal{M}} f \, d\mu(p).$$

$\square$

Theorem 4.3 can be used to identify the number of disconnected structures in very complex structures. Figure 4 shows a part of lung vessel tree obtained from computed tomography (CT) [19, 50]. Gaussian noise is added to one of the coordinates. 3D volumetric heat kernel smoothing with bandwidth 0.01, 0.1, 1 and 10000 is performed on voxels. At the bandwidth 10000, heat kernel smoothing is almost reaching the steady state. Differently colored vessel tree shows they are disconnected structures. There are total 7 disconnected structures.

Since $\psi_0 = 1/\sqrt{\mu(\mathcal{M})}$, we have

$$K_\sigma * f(p) = \frac{\int_{\mathcal{M}} f(p) \, d\mu(p)}{\mu}(\mathcal{M}) + f_1 e^{-\lambda_1 \sigma} \psi_1(p) + R(\sigma, p), \qquad (4.7)$$

where the first term is the average signal, f_1 is a constant and the remaining term R goes to 0 faster than $e^{-\lambda_1 \sigma}$ as $\sigma \to \infty$ [2]. Due to expansion (4.7), the behavior of heat kernel smoothing is basically governed by the second eigenfunction ψ_1 for large bandwidth.

4.2. Inequalities

We are interested in bounding heat kernel smoothing $K_\sigma * f(p)$. We present few useful inequalities involving heat kernel smoothing.

Theorem 4.4: *Conservation of signal:*

$$\int_{\mathcal{M}} K_\sigma * f(p) \, d\mu(p) = \int_{\mathcal{M}} f(p) \, d\mu(p).$$

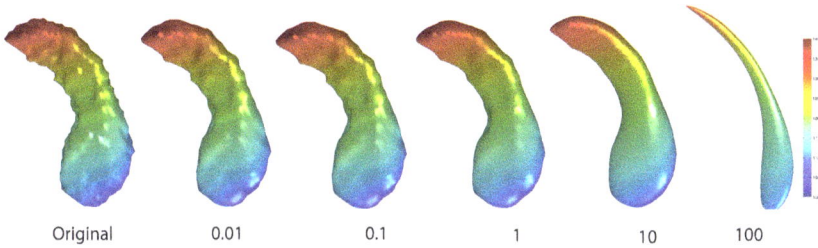

Fig. 5. Heat kernel smoothing on the surface coordinates of hippocampus mesh with different bandwidth.

Proof: This is due to the fact that K_σ is a probability distribution in $\mathcal{M}$, i.e.,

$$\int_{\mathcal{M}} K_\sigma(p,q)\, d\mu(q) = 1$$

for all $p \in \mathcal{M}$. Then

$$\int_{\mathcal{M}} K_\sigma * f(p)\, d\mu(p) = \int_{\mathcal{M}} f(p) \int_{\mathcal{M}} K_\sigma(p,q)\, d\mu(q)\, d\mu(p)$$
$$= \int_{\mathcal{M}} f(p)\, d\mu(p).$$

$\square$

Theorem 4.5: *For $f \in L^1(\mathcal{M})$,*

$$\|K_\sigma * f\|_1 \leq \|f\|_1.$$

Proof: For $f \geq 0$, from Theorem 4.4, trivially we have equality

$$\|K_\sigma * f\|_1 = \|f\|_1.$$

If $f \leq 0$, then $|f| = -f$ and we have the same result. If the sign of f is not constant, consider decomposition $f = f^+ + f^-$, where

$$f^+(p) = \begin{cases} f(p), & f(p) > 0, \\ 0, & f(p) \leq 0. \end{cases} \qquad f^-(p) = \begin{cases} f(p), & f(p) < 0, \\ 0, & f(p) \geq 0. \end{cases}$$

We can write $|f|$ as

$$|f| = |f^+| + |f^-|.$$

Then we have

$$\|K_\sigma * f\|_1 = \int_{\mathcal{M}} \left| K_\sigma * f^+(p) + K_\sigma * f^-(p) \right| d\mu(p) \qquad (4.8)$$

$$\leq \int_{\mathcal{M}} \left| K_\sigma * f^+(p) \right| + \left| K_\sigma * f^-(p) \right| d\mu(p)$$

$$= \|f^+\|_1 + \|f^-\|_1$$

$$= \|f\|_1.$$

$$\square$$

Only when the sign of f does not change, Theorem 4.5 is equality. Consider the following counter example. Consider $\mathcal{M}_1, \mathcal{M}_2 \subseteq \mathcal{M}, \mathcal{M}_1 \cap \mathcal{M}_2 = \emptyset, \mathcal{M}_1 \cup \mathcal{M}_2 = \mathcal{M}$, such that

$$\int_{\mathcal{M}_1} K_\sigma(p,q)\, d\mu(q) = 0.5, \quad \int_{\mathcal{M}_2} K_\sigma(p,q)\, d\mu(q) = 0.5.$$

Set $f(p)$ as

$$f(p) = \begin{cases} 1, & p \in \mathcal{M}_1, \\ -1, & p \in \mathcal{M}_2. \end{cases}$$

Then

$$\|K_\sigma * f\|_1 = \| \int_{\mathcal{M}_1} K_\sigma(p,q)(1)\, d\mu(q) + \int_{\mathcal{M}_2} K_\sigma(p,q)(-1)\, d\mu(q)\|_1$$

$$= \|0.5 + (-0.5)\|_1 = 0.$$

While

$$\|f\|_1 = \int_{\mathcal{M}} 1\, d\mu(p) = \mu(\mathcal{M}).$$

We prove similar result for other L^n-norms.

Theorem 4.6: *For $f \in L^n(\mathcal{M})$, we have norm contraction*

$$\|K_{\sigma_1} * f\|_n \leq \|K_{\sigma_2} * f\|_n \leq \cdots$$

for $\sigma_1 \geq \sigma_2 \geq \cdots \geq 0$ and $n \geq 2$.

Proof: Consider $g \in L^n(\mathcal{M})$. Based on Jensen's inequality and the fact that K_σ is a probability distribution, we have

$$\|K_\sigma * g\|_n^n = \int_{\mathcal{M}} \left| \int_{\mathcal{M}} K_\sigma(p,q)g(q)\, d\mu(q) \right|^n d\mu(p)$$

$$\leq \int_{\mathcal{M}} \int_{\mathcal{M}} K_\sigma(p,q)|g(q)|^n\, d\mu(q)\, d\mu(p)$$

$$= \|K_\sigma * (|g|^n)\|_1 = \|g\|_n^n. \qquad (4.9)$$

Equation (4.9) is due to Theorem 4.5. Now let $\sigma = \sigma_1$ and $g = K_{\sigma_1 - \sigma_2} * f$. Then we have

$$\|K_{\sigma_1} * f\|_n^n = \|K_{\sigma_1 - \sigma_2} * (K_{\sigma_2} * f)\|_n^n \leq \|K_{\sigma_2} * f\|_n^n.$$

$\square$

For L^2-norm, Theorem 4.6 is the consequence of the Hilbert space isomorphism. Note

$$\|K_\sigma * f\|_2 = \|\sum_j e^{-\lambda_j \sigma} f_j \psi_j\|_2 = \sum_{j=0}^{\infty} e^{-2\lambda_j \sigma} f_j^2.$$

Since $\sum_{j=0}^{\infty} e^{-2\lambda_j \sigma} f_j^2 \leq \sum_{j=0}^{\infty} f_j^2$, we have $\|K_\sigma * f\|_2 \leq \|f\|_2$. Theorem 4.6 can be used to bound $K_\sigma * f(p)$ uniformly. From Hölder's inequality and from Theorem 4.6, for each fixed p,

$$\begin{aligned}
|K_\sigma * f(p)| &\leq \int_{\mathcal{M}} K_\sigma(p, q) |f(q)| \, d\mu(q) \\
&\leq \mu(\mathcal{M})^{1/2} \|K_\sigma * f\|_2 \\
&\leq \mu(\mathcal{M})^{1/2} \|f\|_2.
\end{aligned} \tag{4.10}$$

Note we also have $|K_\sigma * f(p)| \leq \sup_{p \in \mathcal{M}} |f(p)|$.

Theorem 4.7: *For $f \in L^\infty(\mathcal{M})$, we have norm contraction*

$$\|K_{\sigma_1} * f\|_\infty \leq \|K_{\sigma_2} * f\|_\infty \leq \cdots$$

for $\sigma_1 \geq \sigma_2 \geq \cdots \geq 0$.

Proof: Consider $g \in L^\infty(\mathcal{M})$. We have

$$\begin{aligned}
|K_\sigma * g(p)| &\leq \int_{\mathcal{M}} K_\sigma(p, q) |g(q)| \, d\mu(q) \\
&\leq \|g\|_\infty \int_{\mathcal{M}} K_\sigma(p, q) \, d\mu(q) \\
&= \|g\|_\infty
\end{aligned}$$

for any $p \in \mathcal{M}$. Hence,

$$\|K_\sigma * g\|_\infty = \sup_{p \in \mathcal{M}} |K_\sigma * g(p)| \leq \|g\|_\infty.$$

Now let $\sigma = \sigma_1$ and $g = K_{\sigma_1 - \sigma_2} * f$. Then we have

$$\|K_{\sigma_1} * f\|_\infty = \|K_{\sigma_1 - \sigma_2} * (K_{\sigma_2} * f)\|_\infty \leq \|K_{\sigma_2} * f\|_\infty.$$

$\square$

The result of smoothing is often used in quantifying a collection of shapes and functions in medical imaging [16, 15]. Thus, we investigate the distance between two different heat kernel smoothings and how it changes from before smoothing. We define L^n-distance between functions $f, g \in L^n(\mathcal{M})$ as that

$$d_n(f, g) = \|f - g\|_n.$$

Then we can show that the distance between functions decreases after heat kernel smoothing.

Theorem 4.8: *Heat kernel smoothing is a contraction map in $L^n(\mathcal{M})$ in a sense that for any $f, g \in L^n(\mathcal{M})$,*

$$d_n(K_{\sigma_1} * f, K_{\sigma_1} * g) \leq d_n(K_{\sigma_2} * f, K_{\sigma_2} * g)$$

for $\sigma_1 \geq \sigma_2 \geq 0$ and $n \geq 1$.

Proof: It follows from Theorems 4.2 (scale-space property) that

$$K_{\sigma_1} * f = K_{\sigma_1 - \sigma_2} * K_{\sigma_2} * f.$$

From Theorem 4.6,

$$\|K_{\sigma_1} * f\|_n = \|K_{\sigma_1 - \sigma_2} * (K_{\sigma_2} * f)\|_n \leq \|K_{\sigma_2} * f\|_n$$

for $\sigma_1 \geq \sigma_2 \geq 0$. □

As a special case of Theorem 4.8, we have

$$d_n(K_\sigma * f, K_\sigma * g) \leq d_n(f, g).$$

Theorem 4.8 also holds true for L^∞-norm as well.

4.3. Discrete heat kernel smoothing on graphs

Discrete heat kernel smoothing of measurement vector $f = (f_1, f_2, \ldots, f_p)^\mathsf{T}$ on a graph is defined similarly as

$$K_\sigma * f = K_\sigma f = \sum_{j=0}^{p} e^{-\lambda_j \sigma} \tilde{f}_j \psi_j. \tag{4.11}$$

This is the discrete analogue of heat kernel smoothing first defined in [17]. In discrete setting, the convolution $*$ is simply a matrix multiplication. Then

$$K_0 * f = f$$

and

$$K_\infty * f = \bar{f}\mathbf{1},$$

where $\bar{f} = \sum_{j=1}^p f_j/p$ is the mean of signal f over every node. When the bandwidth is zero, we are not smoothing data. As the bandwidth increases, the smoothed signal converges to the sample mean of all values. Then we have similar results for the discrete version as well. Here we only show the contraction mapping property simply to illustrate the differences.

Theorem 4.9: *Heat kernel smoothing is a contraction mapping with respect to the l^n-norm for vectors, i.e.,*

$$\|K_\sigma * f\|_n^n \le \|f\|_n^n.$$

Proof: Let kernel matrix $K_\sigma = (k_{ij})$. Then we have inequality

$$\|K_\sigma * f\|_n^n = \sum_{i=1}^p \sum_{j=1}^p |k_{ij} f_j|^n \le \sum_{j=1}^p |f_j|^n.$$

We used Jensen's inequality and doubly-stochastic property of the heat kernel. □

A similar result can be obtained for l^∞-norm.

5. Statistical properties of heat kernel smoothing

Often observed noisy data on graphs is smoothed to increase the signal-to-noise ratio (SNR) and increases the statistical sensitivity [15]. We are interested in knowing how heat kernel smoothing will have effects on the statistical properties on the data. In practice, functional data $Y(p)$ is modeled as a random field:

$$Y(p) = f(p) + e(p), \tag{5.1}$$

where unknown deterministic signal $f \in L^2(\mathcal{M})$ and ϵ is a zero-mean random field with some covariance function $R_e(p, q)$, i.e.,

$$R_e(p, q) = \mathbb{E}[e(p)e(q)].$$

We will further assume constant variance field, i.e.,

$$R_e(p, p) = \mathbb{V}e(p) = \mathbb{E}e^2(p) = \text{const.}$$

for all $p \in \mathcal{M}$.

The covariance functions are often unimodal and isotropic in $\mathcal{M}$. A function is isotropic a manifold in the following sense. Consider line segment $C \subset \mathcal{M}$ connecting p and q and parameterized by $\gamma_c(t)$ with $\gamma_c(0) = p$ and $\gamma_c(1) = q$. In the Cartesian coordinates, $\gamma_c(t) = (\gamma_c^1(t), \ldots, \gamma_c^n(t)) \in \mathbb{R}^n$. The length of C is given by

$$\int_0^1 \langle \frac{d\gamma_c}{dt}, \frac{d\gamma_c}{dt} \rangle^{1/2} \, dt = \int_0^1 \Big[\sum_{i,j} g_{ij} \frac{d\gamma_c^i}{dt} \frac{d\gamma_c^j}{dt} \Big]^{1/2} \, dt$$

where the inner product $\langle \cdot, \cdot \rangle$ is with respect to the tangent space of the manifold. Then the geodesic distance between p and q is defined as the minimizer

$$d_g(p, q) = \min_C \int_0^1 \langle \frac{d\gamma_c}{dt}, \frac{d\gamma_c}{dt} \rangle^{1/2} \, dt.$$

It is usually given as the solution of the Euler equation and numerical techniques are available for polygonal surfaces [49]. Suppose the covariance function of e is of unimodal isotropic function of the form $R_e(p, q) = \rho(d(p, q))$ where $d(p, q)$ is the geodesic distance between p and q and ρ is some non-increasing function. This is an often encountered covariance function shape in applications. Note $d(p, p) = 0$ and $R_e(p, p) = \rho(0)$. Noise $e(p)$ can be further modeled as Gaussian white noise, i.e., Brownian motion or the generalized derivatives of Wiener process, whose covariance function is Dirac-delta, i.e.,

$$R_e(p, q) = \delta(p - q).$$

Often observed functional data $Y(p)$ is smoothed with heat kernel K_σ to increase the signal-to-noise ratio (SNR) and increases the statistical sensitivity [15]. Once heat kernel smoothing is applied to (5.1), we have

$$K_\sigma * Y(p) = K_\sigma * f(p) + K_\sigma * e(p). \tag{5.2}$$

For Gaussian white noise e, its covariance function of $K_\sigma * e$ is given by

$$R_{K_\sigma * e}(p, q) = \int_{\mathcal{M}} K_\sigma(p, r) K_\sigma(q, r) \, d\mu(r).$$

The variance at p is then

$$\mathbb{V}[K_\sigma * e(p)] = R_{K_\sigma * e}(p, p) = \int_{\mathcal{M}} K_\sigma^2(p, r) \, d\mu(r).$$

The variance of data will be often reduced after heat kernel smoothing in the following sense [17, 16]. This is formulated rigorously as follows.

We are interested in determining how the statistical properties of signal change between (5.1) and (5.10). This is needed to study the behavior of collection of heat kernel smoothed functions statistically. We can show that the variance of smoothed noise is smaller than the variance of noise.

Theorem 5.1:

$$\mathbb{V}[K_\sigma * Y(p)] \le \mathbb{V}Y(p)$$

for all $p \in \mathcal{M}$.

Proof: Note that

$$\mathbb{V}e(p) = \mathbb{V}Y(p)$$

$$\mathbb{V}[K_\sigma * Y(p)] = \mathbb{V}[K_\sigma * e(p)].$$

Since $\mathbb{E}(K_\sigma * e(p)) = 0$,

$$\mathbb{V}[K_\sigma * e(p)] = \mathbb{E}\left[\left(K_\sigma * e(p)\right)^2\right].$$

It follows from Theorem 4.4 and Jensen's inequality that

$$\mathbb{E}\left[\int_\mathcal{M} K(p,q)e(q)\,d\mu(q)\right]^2 \le \mathbb{E}\left[\int_\mathcal{M} K(p,q)e(q)^2\,d\mu(q)\right]$$

$$= \mathbb{E}e^2(p)\int_\mathcal{M} K(p,q)\,d\mu(q)$$

$$= \mathbb{E}e^2(p).$$

$\square$

Theorem 5.1 shows heat kernel smoothing reduces the point-wise variability of functional signal. Note the usual t-statistic often used in anatomical shape discrimination analysis [16, 15] is inversely proportional to the standard deviation. Since heat kernel smoothing reduces the variability, t-statistics will likely increase.

5.1. *Heat kernel regression on manifolds*

Consider subspace $\mathcal{H}_k \subset L^2(\mathcal{M})$ spanned by the orthonormal basis $\{\psi_j\}$, i.e.,

$$\mathcal{H}_k = \{\sum_{j=0}^{k} \beta_j \psi_j(p) : \beta_j \in \mathbb{R}\}.$$

Then the least squares estimation (LSE) of unknown signal f in $\mathcal{H}_k$ in model (5.1) is given by the shortest distance from observed signal Y to $\mathcal{H}_k$:

$$\widehat{f}(p) = \arg\min_{h \in \mathcal{H}_k} \int_{\mathcal{M}} |Y(p) - h(p)|^2 \, d\mu(p) = \sum_{j=0}^{k} Y_j \psi_j(p), \qquad (5.3)$$

where $Y_j = \langle Y, \psi_j \rangle$ are the Fourier coefficients. This is the usual Fourier series expansion that tends to suffer the Gibbs phenomenon, i.e., ringing artifact, for compact surfaces [13, 22]. The Gibbs phenomenon can be effectively removed if the Fourier series expansion converges fast enough as the number of basis functions goes to infinity. By weighting the Fourier coefficients exponentially smaller, we can make the representation converges faster; this can be achieved by additionally weighting the squared residuals in equation (5.3) with the heat kernel:

$$\widehat{f}(p) = \arg\min_{h \in \mathcal{H}_k} \int_{\mathcal{M}} \int_{\mathcal{M}} K_\sigma(p, q) |Y(q) - h(p)|^2 \, d\mu(q) \, d\mu(p). \qquad (5.4)$$

The optimization (5.4) has the following analytic expression:

Theorem 5.2:

$$\widehat{f}(p) = \arg\min_{h \in \mathcal{H}_k} \int_{\mathcal{M}} \int_{\mathcal{M}} K(p, q) |Y(q) - h(p)|^2 \, d\mu(q) \, d\mu(p) = \sum_{j=0}^{k} \tau_j Y_j \psi_j,$$

where $Y_j = \langle Y, \psi_j \rangle$ are Fourier coefficients.

Proof: Any function $h \in \mathcal{H}_k$ can be expressed as

$$h(p) = \sum_{j=0}^{k} \beta_j \psi_j(p). \qquad (5.5)$$

Then by plugging (5.5) into the inner integral $I(p)$, it becomes

$$I(p) = \int_{\mathcal{M}} K_\sigma(p, q) \left| Y(q) - \sum_{j=0}^{k} \beta_j \psi_j(p) \right|^2 \, d\mu(q).$$

Simplifying the expression, we obtain

$$I(p) = \sum_{j=0}^{k} \sum_{j'=0}^{k} \psi_j(p)\psi_{j'}(p)\beta_j\beta_{j'} - 2K_\sigma * Y(p) \sum_{j=0}^{k} \psi_j(p)\beta_j + K_\sigma * Y^2(p).$$

$$(5.6)$$

The kernel can be written as

$$K_\sigma(p,q) = \sum_{j'=0}^{\infty} \tau_{j'} \psi_{j'}(p) \psi_{j'}(q) \qquad (5.7)$$

and the convolution is written as

$$K_\sigma * Y(p) = \sum_{j'=0}^{\infty} \tau_{j'} Y_{j'} \psi_{j'}(p).$$

Since I is an unconstrained positive semidefinite quadratic program (QP) in β_j, there is no unique global minimizer of I without additional linear constraints. Integrating I further with respect to $d\mu(p)$, we collapse (5.6) to a positive definite QP, which yields a unique global minimizer:

$$\int_{\mathcal{M}} I(p)\, d\mu(p) = \sum_{j=0}^{k} \beta_j^2 - 2\sum_{j=0}^{k} \tau_j Y_j \beta_j + \text{ const.}$$

The minimum of the above integral is obtained when all the partial derivatives with respect to β_j vanish, i.e.

$$\int_{\mathcal{M}} \frac{\partial I}{\partial \beta_j}\, d\mu(p) = 2\beta_j - 2\tau_j Y_j = 0$$

for all j. Hence $\sum_{j=0}^{k} \tau_j Y_j \psi_j$ must be the unique minimizer. $\qquad \square$

Theorem 5.2 generalizes the weighted spherical harmonic (SPHARM) representation on a unit sphere to an arbitrary manifold [14]. Theorem 5.2 implies that the kernel regression can be performed by simply computing the Fourier coefficients $f_j = \langle f, \psi_j \rangle$ without doing any numerical optimization. The numerically difficult optimization problem is reduced to the problem of computing Fourier coefficients. If the kernel K is a Dirac-delta function, the kernel regression simply collapses to the least squares estimation (LSE) which results in the standard Fourier series, i.e.

$$\widehat{f}(p) = \arg\min_{h \in \mathcal{H}_k} \int_{\mathcal{M}} \left| Y(q) - h(q) \right|^2 d\mu(q) = \sum_{j=0}^{k} f_j \psi_j.$$

It can be also shown that as $k \to \infty$, the kernel regression

$$\widehat{f} = \sum_{j=0}^{k} \tau_j Y_j \psi_j$$

converges to convolution $K_\sigma * Y$ establishing the connection to the manifold-based kernel smoothing framework [3, 17]. Hence, asymptotically the proposed kernel regression should inherit many statistical properties of kernel smoothing.

5.2. Statistical properties on graphs

Similar results can be obtained for graph data structures. Consider the following model:

$$f = \mu + e,$$

where μ is an unknown signal and ϵ is zero mean noise. Let $e = (e_1, \dots, e_p)^\mathsf{T}$. Denote $\mathbb{E}$ as expectation and $\mathbb{V}$ as covariance. It is natural to assume that the variability of noises at different nodes j is identical, i.e.,

$$\mathbb{E}e_1^2 = \mathbb{E}e_2^2 = \cdots = \mathbb{E}e_p^2. \tag{5.8}$$

Further, we assume that data at two nodes i and j to have less correlation when the distance between the nodes is large. So covariance matrix $R_e = \mathbb{V}e = \mathbb{E}(ee^\mathsf{T}) = (r_{ij})$ can be given by

$$r_{ij} = \rho(d_{ij}) \tag{5.9}$$

for some decreasing function ρ and geodesic distance d_{ij} between nodes i and j. Note $r_{jj} = \rho(0)$ with the understanding that $d_{jj} = 0$ for all j. The off diagonal entries of R_e are smaller than the diagonal. Noise e can be further modeled as the discrete Gaussian white noise whose covariance matrix elements are Kroneker-delta δ_{ij} with $\delta_{ij} = 1$ if $i = j$ and 0 otherwise. Thus,

$$R_e = \mathbb{E}(ee^\mathsf{T}) = I_p,$$

the identity matrix of size $p \times p$. Since $\delta_{jj} \geq \delta_{ij}$, Gaussian white noise is a special case of (5.9). After heat kernel smoothing, we have

$$K_\sigma * f = K_\sigma * \mu + K_\sigma * e. \tag{5.10}$$

For $R_e = I_p$, the covariance matrix of smoothed noise is simply given as

$$R_{K_\sigma * e} = K_\sigma \mathbb{E}(ee^\mathsf{T}) K_\sigma = K_\sigma^2 = K_{2\sigma}.$$

We used the scale-space property of heat kernel. In general, the covariance matrix of smoothed data $K_\sigma * e$ is given by

$$R_{K_\sigma * e} = K_\sigma \mathbb{E}(ee^\mathsf{T}) K_\sigma = K_\sigma R_e K_\sigma.$$

Other than these differences, similar analogous results can be obtained.

5.3. *Persistent homology in heat kernel smoothing*

In persistent homology, a point cloud is used to build a Rips filtration [1,7,9]. Similarly, we can build Rips filtration in a function space. Given a collection of functional measurements in $L^n(\mathcal{M})$, heat kernel smoothing induces a Rips filtration in $L^n(\mathcal{M})$ if we take the functions as a point cloud and build a filtration using L^n-norm as distance.

Theorem 5.3: *Let $A_\sigma = \{f \in L^n(\mathcal{M}) : \|K_\sigma * f\|_n \le h\}$. Then A_σ induces filtration*

$$A_{\sigma_1} \subset A_{\sigma_2} \subset \cdots$$

for any $\sigma_1 \ge \sigma_2 \ge \cdots \ge 0$, $h \ge 0$ and $n \ge 1$.

Proof: Suppose $f \in A_{\sigma_1}$. From Theorem 4.6 (norm contraction),

$$\|K_{\sigma_1} * f\|_n \le \|K_{\sigma_2} * f\|_n \le h.$$

Then $f \in A_{\sigma_2}$ and the result follows. $\qquad\square$

A similar result can be obtained for $L^\infty(\mathcal{M})$ space as well. Theorem 5.3 build filtrations on the space of functions. We can also build a filtration directly in manifold $\mathcal{M}$ as well.

Theorem 5.4: *Let $B_\sigma = \{p \in \mathcal{M} : \mathbb{V}[K_\sigma * Y(p)] \le h\}$. Then B_σ satisfies*

$$B_{\sigma_1} \subset B_{\sigma_2} \subset \cdots B_0 \tag{5.11}$$

if $\sigma_1 \ge \sigma_2 \ge \cdots \ge 0$ for any $h \ge 0$ and $n \ge 1$.

Proof: Let $p \in B_{\sigma_1}$. Then from Theorem 5.1,

$$\mathbb{V}[K_{\sigma_1} * Y(p)] = \mathbb{V}[K_{\sigma_1 - \sigma_2} * (K_{\sigma_2} * Y)(p)]$$
$$\le \mathbb{V}[K_{\sigma_2} * Y(p)] \le h.$$

Thus, $p \in B_{\sigma_2}$. $\qquad\square$

6. Discussion

For irregular domains in images, boundary shapes are often complex. This causes the geometric shape of the boundary to strongly bias smoothing. [43] proposed more natural boundary conditions that reduces the boundary induced bias in smoothing by using the Neumann boundary condition in solving a partial different equation. The heat kernel smoothing method proposed here is based on the Dirichlet boundary condition although extending it to the Neumann boundary condition is also possible. For closed surfaces with no boundary, there is no need to consider for the boundary condition.

Acknowledgment

This work was partially supported by the NIH research grant R01 EB022856. We would like to thank Jim Ramsay of McGill University, Michelle Carey of University College Dublin and Yu-Min Chung of College of Willam and Mary for valuable discussions on heat kernel smoothing in irregular domains. We would like to thank Ruth Sullivan, Michael Johnson and Michael Newton of University of Wisconsin-Madison for useful discussions on modeling vessel trees. Gurong Wu of University of North Carolina-Chapel Hill provided the lung vessel tree data illustrated in this chapter.

References

1. R.J. Adler, O. Bobrowski, M.S. Borman, E. Subag, and S. Weinberger. Persistent homology for random fields and complexes. In *Borrowing strength: theory powering applications–a Festschrift for Lawrence D. Brown*, pages 124–143. Institute of Mathematical Statistics, 2010.
2. R. Banuelos and K. Burdzy. On the Hot Spots Conjecture of J. Rauch. *Journal of Functional Analysis*, 164:1–33, 1999.
3. M. Belkin and P. Niyogi. Laplacian eigenmaps and spectral techniques for embedding and clustering. In *Advances in Neural Information Processing Systems*, pages 585–592, 2002.
4. M. Belkin, P. Niyogi, and V. Sindhwani. Manifold regularization: A geometric framework for learning from labeled and unlabeled examples. *The Journal of Machine Learning Research*, 7:2399–2434, 2006.
5. N. Berline, E. Getzler, and M. Vergne. *Heat kernels and Dirac operators*. Springer-Verlag, 1991.
6. M.M. Bronstein and I. Kokkinos. Scale-invariant heat kernel signatures for non-rigid shape recognition. In *IEEE Conference on Computer Vision and Pattern Recognition (CVPR)*, pages 1704–1711, 2010.
7. P. Bubenik and P.T. Kim. A statistical approach to persistent homology. *Homology Homotopy and Applications*, 9:337–362, 2007.
8. A. Bueno-Orovio. Fourier embedded domain methods: periodic and c^∞ extension of a function defined on an irregular region to a rectangle via convolution with Gaussian kernels. *Applied Mathematics and Computation*, 183:813–818, 2006.
9. G. Carlsson and F. Memoli. Persistent clustering and a theorem of J. Kleinberg. *arXiv preprint arXiv:0808.2241*, 2008.

10. F.R.K. Chung and S.T. Yau. Eigenvalue inequalities for graphs and convex subgraphs. *Communications in Analysis and Geometry*, 5:575–624, 1997.

11. M.K. Chung. *Statistical Morphometry in Neuroanatomy*. Ph.D. Thesis, McGill University, 2001. http://www.stat.wisc.edu/~mchung/papers/thesis.pdf.

12. M.K. Chung, K.M. Dalton, and R.J. Davidson. Tensor-based cortical surface morphometry via weighted spherical harmonic representation. *IEEE Transactions on Medical Imaging*, 27:1143–1151, 2008.

13. M.K. Chung, K.M. Dalton, L. Shen, A.C. Evans, and R.J. Davidson. Weighted Fourier representation and its application to quantifying the amount of gray matter. *IEEE Transactions on Medical Imaging*, 26:566–581, 2007.

14. M.K. Chung, R. Hartley, K.M. Dalton, and R.J. Davidson. Encoding cortical surface by spherical harmonics. *Statistica Sinica*, 18:1269–1291, 2008.

15. M.K. Chung, A. Qiu, S. Seo, and H.K. Vorperian. Unified heat kernel regression for diffusion, kernel smoothing and wavelets on manifolds and its application to mandible growth modeling in CT images. *Medical Image Analysis*, 22:63–76, 2015.

16. M.K. Chung, S. Robbins, K.M. Dalton, R.J. Davidson, A.L. Alexander, and A.C. Evans. Cortical thickness analysis in autism with heat kernel smoothing. *NeuroImage*, 25:1256–1265, 2005.

17. M.K. Chung, S. Robbins, and A.C. Evans. Unified statistical approach to cortical thickness analysis. *Information Processing in Medical Imaging (IPMI)*, *Lecture Notes in Computer Science*, 3565:627–638, 2005.

18. M.K. Chung and J. Taylor. Diffusion smoothing on brain surface via finite element method. In *Proceedings of IEEE International Symposium on Biomedical Imaging (ISBI)*, volume 1, pages 432–435, 2004.

19. M.K. Chung, Y. Wang, and G. 2018 Wu. Heat kernel smoothing in irregular image domains. *International Conference of the IEEE Engineering in Medicine and Biology Society (EMBC)*, 2018.

20. M. Desbrun, E. Kanso, and Y. Tong. Discrete differential forms for computational modeling. In *Discrete differential geometry*, pages 287–324. Springer, 2008.

21. G. Domokos. Four-dimensional symmetry. *Physical Review*, 159:1387–1403, 1967.

22. A. Gelb. The resolution of the Gibbs phenomenon for spherical harmonics. *Mathematics of Computation*, 66:699–717, 1997.

23. G.M.L. Gladwell and H. Zhu. Courant's nodal line theorem and its discrete counterparts. *The Quarterly Journal of Mechanics and Applied Mathematics*, 55:1–15, 2002.

24. L.R. Haff, P.T. Kim, J.-Y. Koo, and D.S.P. Richards. Minimax estimation for mixtures of wishart distributions. *The Annals of Statistics*, 39:3417–3440, 2011.

25. A.P. Hosseinbor, M.K. Chung, C.G. Koay, S.M. Schaefer, C.M. Van Reekum, L.P. Schmitz, M. Sutterer, A.L. Alexander, and R.J. Davidson. 4D hyperspherical harmonic (HyperSPHARM) representation of surface anatomy: A

holistic treatment of multiple disconnected anatomical structures. *Medical Image Analysis*, 22:89–101, 2015.

26. A.P. Hosseinbor, M.K. Chung, Y.-C. Wu, B.B. Bendlin, and A.L. Alexander. A 4D hyperspherical interpretation of q-space. *Medical Image Analysis*, 21:15–28, 2015.

27. A.P. Hosseinbor, W.H. Kim, N. Adluru, A. Acharya, H.K. Vorperian, and M.K. Chung. The 4D hyperspherical diffusion wavelet: A new method for the detection of localized anatomical variation. In *International Conference on Medical Image Computing and Computer-Assisted Intervention*, volume 8675, pages 65–72, 2014.

28. A.T. James. Calculation of zonal polynomial coefficients by use of the Laplace-Beltrami operator. *The Annals of Mathematical Statistics*, 39:1711–1718, 1968.

29. E. Kreyszig. *Differential Geometry*. University of Toronto Press, 1959.

30. B. Lévy. Laplace-Beltrami eigenfunctions towards an algorithm that "understands" geometry. In *IEEE International Conference on Shape Modeling and Applications*, page 13, 2006.

31. C. Li and V. Alexiades. Time stepping for the cable equation, Part 1: Serial performance. *Proceedings of Neural, Parallel & Scientific Computations*, 4:241–246, 2010.

32. H. Maass. Die bestimmung der dirichletreihen mit grössencharakteren zu den modulformen n-ten grades. *Journal of Indian Mathematical Society*, 19:1–23, 1955.

33. J.E. Marsden and T.J.R. Hughes. *Mathematical Foundations of Elasticity*. Dover Publications, Inc., 1983.

34. J. Nilsson, F. Sha, and M.I. Jordan. Regression on manifolds using kernel dimension reduction. In *Proceedings of the 24th International Conference on Machine Learning*, pages 697–704. ACM, 2007.

35. A. Qiu, D. Bitouk, and M.I. Miller. Smooth functional and structural maps on the neocortex via orthonormal bases of the Laplace-Beltrami operator. *IEEE Transactions on Medical Imaging*, 25:1296–1396, 2006.

36. D.S.P. Richards. Applications of invariant differential operators to multivariate distribution theory. *SIAM Journal on Applied Mathematics*, 45:280–288, 1985.

37. D.S.P. Richards. High-dimensional random matrices from the classical matrix groups, and generalized hypergeometric functions of matrix argument. *Symmetry*, 3:600–610, 2011.

38. S. Rosenberg. *The Laplacian on a Riemannian Manifold*. Cambridge University Press, 1997.

39. B. Schölkopf and A.J. Smola. *Learning with Kernels: Support Vector Machines, Regularization, Optimization, and Beyond*. MIT Press, 2002.

40. S. Seo, M.K. Chung, and H.K. Vorperian. Heat kernel smoothing using Laplace-Beltrami eigenfunctions. In *Medical Image Computing and Computer-Assisted Intervention — MICCAI 2010*, volume 6363 of *Lecture Notes in Computer Science*, pages 505–512, 2010.

41. J. Shawe-Taylor and N. Cristianini. *Kernel methods for pattern analysis.* Cambridge University Press, 2004.
42. D.I. Shuman, S.K. Narang, P. Frossard, A. Ortega, and P. Vandergheynst. The emerging field of signal processing on graphs: Extending high-dimensional data analysis to networks and other irregular domains. *IEEE Signal Processing Magazine*, 30:83–98, 2013.
43. O. Stein, E. Grinspun, M. Wardetzky, and A. Jacobson. Natural boundary conditions for smoothing in geometry processing. *arXiv preprint arXiv:1707.04348*, 2017.
44. Florian Steinke and Matthias Hein. Non-parametric regression between manifolds. *Advances in Neural Information Processing Systems*, 2008.
45. J. Sun, M. Ovsjanikov, and L. J. Guibas. A concise and provably informative multi-scale signature based on heat diffusion. *Comput. Graph. Forum*, 28:1383–1392, 2009.
46. T. Tlusty. A relation between the multiplicity of the second eigenvalue of a graph laplacian, courants nodal line theorem and the substantial dimension of tight polyhedral surfaces. *Electrnoic Journal of Linear Algebra*, 16:315–24, 2007.
47. G. Wang, X. Zhang, Q. Su, J. C., J. Chen, L. Wang, Y. Ma, Q. Liu, L. Xu, J. Shi, and Y. Wang. A heat kernel based cortical thickness estimation algorithm. In *International Workshop on Multimodal Brain Image Analysis*, pages 233–245, 2013.
48. G. Wang, X. Zhang, Q. Su, J. Shi, R.J. Caselli, Y. Wang, and Alzheimer's Disease NeuroImaging Initiative. A novel cortical thickness estimation method based on volumetric Laplace-Beltrami operator and heat kernel. *Medical Image Analysis*, 22:1–20, 2015.
49. E. Wolfson and E.L. Schwartz. Computing minimal distances on polyhedral surfaces. *IEEE Transactions on Pattern Analysis and Machine Intelligence*, 11:1001–1005, 1989.
50. G. Wu, Q. Wang, J. Lian, and D. Shen. Estimating the 4d respiratory lung motion by spatiotemporal registration and super-resolution image reconstruction. *Medical Physics*, 40(3):031710, 2013.
51. F. Yger and A. Rakotomamonjy. Wavelet kernel learning. *Pattern Recognition*, 44(10-11):2614–2629, 2011.